Expeditions in Y[...] Geometry

for Common Core State Standards

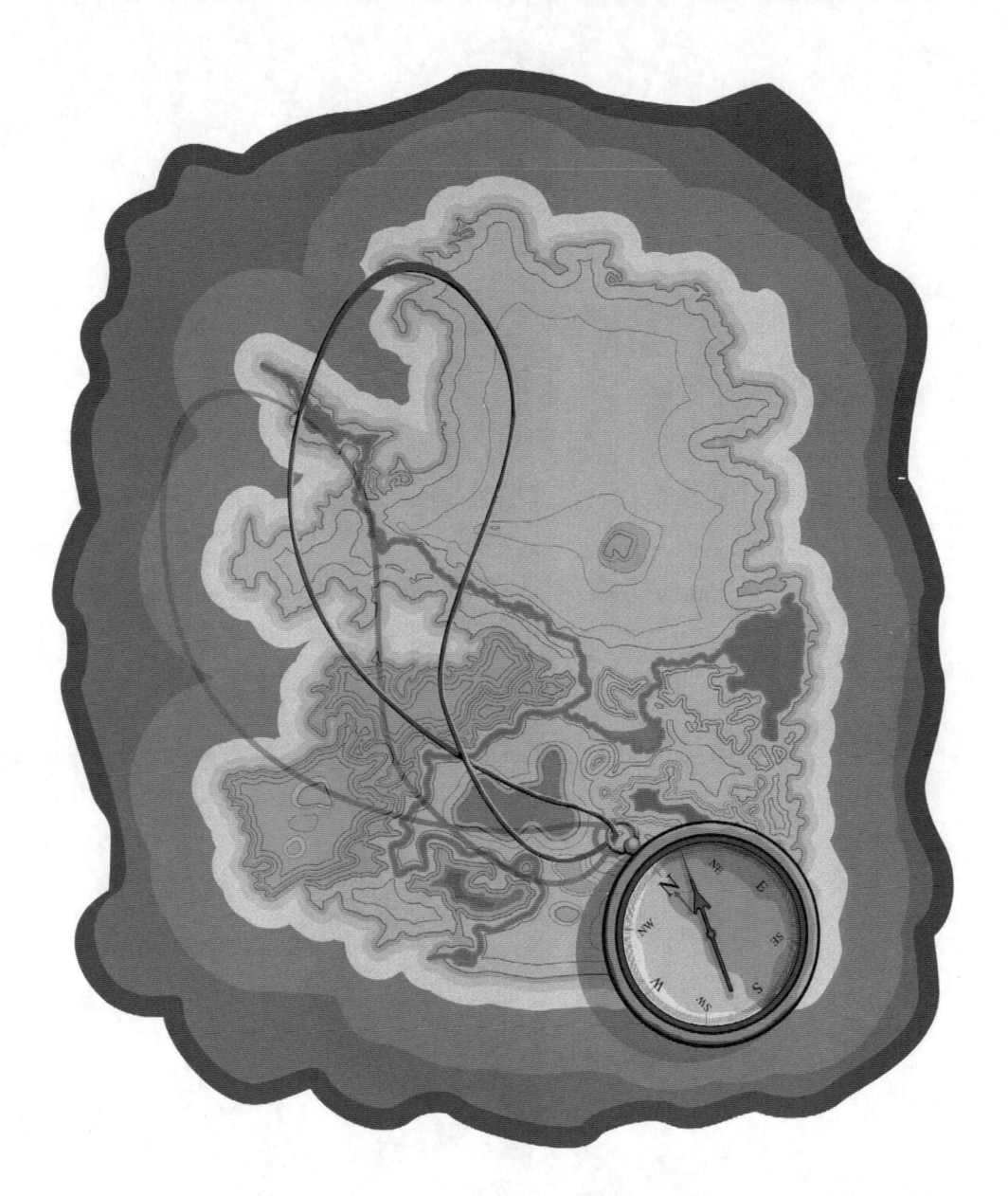

Nora Priest

WALCH EDUCATION

The classroom teacher may reproduce materials in this book for classroom use only.
The reproduction of any part for an entire school or school system is strictly prohibited.
No part of this publication may be transmitted, stored, or recorded in any form
without written permission from the publisher.

© Common Core State Standards. Copyright 2010. National Governor's Association Center for
Best Practices and Council of Chief State School Officers. All rights reserved.

1 2 3 4 5 6 7 8 9 10
ISBN 978-0-8251-6897-0
Copyright © 2007, 2012
J. Weston Walch, Publisher
40 Walch Drive • Portland, ME 04103
www.walch.com
Printed in the United States of America

WALCH EDUCATION®

Contents

Introduction .... *v*

Project Skills Chart .... *vii*

Geometry Project Assessment Rubric .... *viii*

Project Putt-Putt 1

Ripping Rooms 18

Fashionistas 40

At the Scene of the Crime 64

Protectors of the Realm 95

Superhero Challenge 113

Thinking Outside the Box 130

Director's View 151

This Is Air Traffic Control 168

The Great Geometry Race 195

Introduction

We all remember a project we did in school, often with more vivid recall than we can summon for entire courses or years. And for good reason. Projects command attention. They force students to grapple with new information, skills, and technologies in ways that embed learning in memory. They contextualize education and help students truly understand why "I need to know that."

This book contains ten projects designed to leave a lasting mark. These projects provide students with authentic tasks involving real problems, real products, and real people, and use themes that hook young people. At the same time, they have teachers thoroughly in mind.

The high-school curriculum is packed, and, as teachers well know, a project can quickly take on a life of its own. *Expeditions in Your Classroom: Geometry for Common Core State Standards* provides activities and materials that scaffold student tasks, set clear criteria for final products, and offer assessment tools and a detailed outline of project steps so that teachers can focus energy on instruction rather than project management. Each expedition addresses the Common Core State Standards and provides accessible routes to understanding for a broad audience of students.

About Project-Based Learning

In *Real Learning, Real Work*[1], Adria Steinberg describes the qualities of powerful projects: the six A's.

Authenticity
Students solve problems and questions that are meaningful and real. People outside school walls tackle the same challenges. What students create and do has value beyond school.

Academic Rigor
Students encounter challenging material and learn critical skills, knowledge, and habits of mind essential for success in one or more disciplines.

Applied Learning
Students put their knowledge and skills to work in hands-on ways, and learn how to organize and manage themselves along the way.

Active Exploration
Students go into the field. They investigate and communicate their discoveries.

Adult Relationships
Students connect with adults with relevant expertise. They observe them, work with them, and get support and feedback.

Assessment
Students play an active role in defining their goals and assessing their progress. Adults around them give them ongoing and varied opportunities to demonstrate progress.

[1] Steinberg, Adria. *Real Learning, Real Work (Transforming Teaching)*. New York, NY: Routledge, 1998.

Introduction

Project Format and Materials
Each project contains the following materials:

Teacher Pages
- Overview: information on project learning goals, prior knowledge or experience needed by students, time needed for the project, and team formation information
- Suggested Steps: a day-by-day view of how to deliver project activities
- Project Management Tips and Notes: suggestions for how to handle possible issues or information on project options and variations
- Extension Activities: suggested activities for extending the project or exploring related areas
- Common Core State Standards Connection: a list of standards students will address through the project
- Answer Key: answers for Before You Go and Skill Check questions (Many answers will vary, and therefore, have been omitted from the answer keys.)

Student Pages
- Expedition Overview: a description of the project challenge, learning objectives, key vocabulary terms, materials needed, and web resources students use for project activities
- Before You Go: lead-in activities designed to review fundamental skills or knowledge needed for the project
- Off You Go: activities that support the core project, including guidelines and instructions for final products or presentations
- Expedition Tools: handouts and worksheets associated with project activities
- Check Yourself: two assessment tools that students use to check skill development (practice problems or questions) and evaluate their project performance overall

A Geometry Project Assessment Rubric is also included and can be used with any project.

Project Skills Chart

Projects challenge students to flex more than one mental muscle at a time and integrate skills they often see dissected and covered in discrete math book chapters. Each project in this book has a core skill focus, but also gives students an opportunity to practice other skills. Use this chart as a reference to help you find the best project for your needs.

C = Core skill

X = Other skills covered (sometimes optional)

Project	Page	Measurement	Ratio and proportion	Scale drawing	Classifying polygons	Triangle and angle measurements	Polygon measurements	Circle geometry	Three-dimensional shapes and visualization	Transformations	Coordinate systems	Calculating slope	Geometric modeling	Graph theory	Basic trigonometry
Project Putt-Putt	1	X	X	C		C	X		X	X		X			
Ripping Rooms	18	X	X	C	C	X	C	X							
Fashionistas	40	X	C	X			X		C						
At the Scene of the Crime	64	X	X	C		C					C				
Protectors of the Realm	95	X			X	X					C		C		
Superhero Challenge	113	X		X		X					X		C	C	
Thinking Outside the Box	130	X	X	X		X		X	C	X					
Director's View	151	X		X		X		C							C
This Is Air Traffic Control	168	X		X		X				X	C				C
The Great Geometry Race	195	X	X		X	X	X	X	X	X	X	X			

Expeditions in Your Classroom: Geometry © Walch Education

Geometry Project Assessment Rubric

	Percent of grade	4 (Excellent)	3 (Good)	2 (Fair)	1 (Poor)
Knowledge and skills specific to project		Defines all key vocabulary, with examples. Recalls all formulas and methods correctly; can explain and apply to other problems. If required, work shows evidence of research on topic or theme.	Defines majority of terms, with examples. Majority of formulas and methods applied correctly. Can apply to other problems with some incorrect answers. Shows evidence of research.	Definitions and explanations are confusing or incorrect. Some formulas are used correctly. Shows little evidence of research.	No knowledge evident. There are few correct methods and few correct answers. There is no evidence of research.
Measurement/ calculations		Uses correct formulas. Includes all calculations and diagrams used for solution. Answers are correct.	Majority of formulas are correct. Most work is shown. There are some incorrect answers.	Some formulas are used correctly. Some work is shown. There are a number of incorrect answers.	There are few correct formulas, little work shown, and a small number of correct answers.
Drawing and modeling		Final work meets criteria and exceeds expectations. All elements are included and correctly labeled. Work shows mastery of technique/technical skill. If required, scale and proportion are represented accurately.	Final work meets criteria. Majority of elements are included and labeled. Work shows good command of technique/technical skill. If required, scale and proportion are represented accurately.	Final work is missing important elements. Technique is weak. Scale and proportion are not represented accurately.	Did not do work/contribute. Did not attempt to learn technique.
Level of challenge		Investigated difficult or complex situations.	Investigated moderately difficult situations.	Investigated straightforward situations.	Investigated simplest or easiest situations.
Final product		Meets all criteria. Organization and information exceed expectations. Work reflects excellent understanding of project content.	Meets all criteria. Organization and information are presented clearly. Work reflects good understanding of project content.	Meets most criteria. Some elements or components are missing.	Did not contribute. Did not submit or is missing major components.

Expeditions in Your Classroom: Geometry © Walch Education

Geometry Project Assessment Rubric, *continued*

	Percent of grade	4 (Excellent)	3 (Good)	2 (Fair)	1 (Poor)
Presentation		Completed within specific time. Evidence of preparation is obvious. Emphasized most important information. All team members were involved.	Almost completed within time. Some preparation is evident. Covers majority of main points. Not all team members were involved.	Almost completed within time. Little preparation is evident. Misses a number of important points. Not all team members were involved.	Did not participate, did not prepare, was way under or over time, or information was confusing and disjointed.
Teamwork		Workload was divided and shared equally by all members.	Most members, including student, contributed fair share.	Workloads varied considerably. Student did not contribute fair share.	Few members contributed. Student made little to no contribution.
Class participation		Contributed substantially.	Contributed fair share.	Contributed some.	Contributed very little.

Expeditions in Your Classroom: Geometry

Project Putt-Putt

Overview
Students design a miniature golf course. They create blueprints and a model for a new championship course.

Time
Total time: 6 to 8 hours

- Before You Go—Reflection Inspection: 30 to 55 minutes
- Before You Go—Uphill and Downhill: 15 minutes
- Activity 1—Putt-Putt Blueprints: two to four 55-minute class periods
- Concept sketch development: one to two 55-minute class periods
- Scale drawings: one to two 55-minute class periods and one to two hours of homework
- Activity 2—Mini Model: one to two 55-minute class periods and one to two hours of homework

Skill Focus
- angles and reflection
- slope
- two- and three-dimensional modeling

Prior Knowledge
- measurement
- drawing to scale
- basic understanding of angles and their properties

Team Formation
Students can work individually or in teams of two or three students.

Lingo to Learn—Terms to Know
- **angle of incidence:** the angle that a line makes with a line perpendicular to the surface at the point of incidence
- **angle of reflection:** angle measurement between a reflected ray and a line perpendicular to the reflecting surface/line at the point of incidence
- **area:** the number of square units needed to cover a surface
- **congruence:** when figures or angles have the same size and shape
- **isometry:** symmetry; a transformation that is a reflection or a composite of reflections (reflections, rotations, translations/slides, and glide reflections are isometries)

Expeditions in Your Classroom: Geometry

Project Putt-Putt

- **line of reflection:** a line used to create a reflection of a shape (reflecting line or mirror)
- **perimeter:** the sum of the lengths of the sides of a polygon
- **ratio:** a pair of numbers that compares different types of units
- **reflection:** a transformation resulting from a flip
- **scale drawing:** a drawing that is a reduction or an enlargement of the original
- **slope:** the steepness of a line; the measure of change in a surface value over distance

Suggested Steps

Preparation

- Review the list of materials and collect anything you will provide (a golf ball and marble for each team, art supplies, and so forth).
- Review the Miniature Golf Course Hole Specifications in Activity 1. Adapt specifications to fit your skill focus. For example, you may wish to make sloped elements optional.

Day 1

1. Provide an overview of the project and review materials.

2. Facilitate Before You Go: Reflection Inspection, which addresses angles of reflection. Let students practice and discuss observations. Show how to calculate angles of incidence and reflection using the tangent function.

3. Introduce Before You Go: Uphill and Downhill, an activity on how to determine the slope and angle of slope of an incline.

Homework

If students have Internet access at home, have them use the Helpful Web Resources to learn about golf course design and look for examples of miniature golf courses. Encourage them to visit sites that highlight interesting or famous miniature golf courses.

Day 2

1. Explain Activity 1: Putt-Putt Blueprints.

2. Review course holes and drawing specifications. Provide any specific criteria you have (for example, if you do not want students to include slope).

3. If this is not an individual project, ask students to select a partner or assign groups.

Project Putt-Putt

4. Provide due dates for assignments (freehand sketches, scale drawings) and/or specify whether class time will be used.

5. Give students a refresher on ratio and scale if needed.

6. Allow students to begin brainstorming and planning course hole ideas.

Homework

Have students continue to brainstorm ideas for their course holes and create a freehand sketch.

Days 3 through 5

1. Give students time to work on drawings. Alternatively, if done as homework, ask students to show signs of progress each day.

2. Invite students to describe their ideas, show a first draft or revised sketch, and so forth.

3. Check with students to make sure they are on target. Clarify any misconceptions.

Design Due Date

1. Have students hang design drawings around the classroom.

2. Allow 5 to 10 minutes for viewing.

3. Solicit observations and feedback. Discuss any design or construction challenges evident in drawings.

4. Explain Activity 2: Mini Model. Review 3-D model criteria. Add other criteria appropriate to your situation and supplies.

5. Assign a due date for models.

Model Due Date

1. Position course holes on desks around the classroom. Number each hole.

2. Give students wooden craft sticks and marbles.

3. Let students play a few rounds. You might also have them track scores on improvised score cards—notebook paper with three columns for hole number, hole par, and score.

Expeditions in Your Classroom: Geometry © Walch Education

Project Putt-Putt

4. Evaluate the course. Discussion questions might include:
   - Did anyone get a hole in one?
   - Which course holes were most challenging and why? Which course holes were least challenging?
   - Does understanding reflection help your game?

Final Day

1. Have students complete the Skill Check problems.
2. Check and review answers.
3. Have students complete the Self-Assessment and Reflection worksheet and submit it (optional).

Project Management Tips and Notes

- Review proposed designs. Student course holes tend to get complex! Some students may need redirection to simplify the design. Others may need a gentle reminder that the geometry of the course hole and how it plays are more important than appearance.
- As written, the project specifies that course holes be as big as a student's desk and no bigger than two or three desks. You may want to limit designs to one desktop. This is great for practical reasons (supplies, space); however, students may find the space tight for a course hole that should include two bounces and a slope. You can also give a range of dimensions (for instance, larger than 2 feet by 2 feet but smaller than 4 feet by 4 feet). The space doesn't need to be square.

Suggested Assessment

Use the Geometry Project Assessment Rubric or the following point system:

Team and class participation	15 points
Two scaled hole drawings	40 points
3-D course hole model	40 points
Project self-assessment	5 points

Project Putt-Putt

Extension Activities
- Technologically inclined students may use The Geometer's Sketchpad (www.dynamicgeometry.com) or other tools to calculate and model angles before creating sketches.
- Consider having students use a computer-aided design (CAD) program to draft a blueprint or three-dimensional model of their course hole.
- Explore other angles involved in golf: a golf swing, golf clubs, the position of course holes in relation to one another, and so forth.
- Ask students to identify and price the materials they need to build their course hole.
- As a class, design and build a real miniature golf course. Hold a tournament involving local leaders. Use the event as a fund-raiser. (See Junkyard Golf & Potluck: http://junkyardsports.com/events/golfest.pdf)

Common Core State Standards Connection
High School
Geometry: Modeling with Geometry

G-MG.1. Use geometric shapes, their measures, and their properties to describe objects (e.g., modeling a tree trunk or a human torso as a cylinder).★

G-MG.3. Apply geometric methods to solve design problems (e.g., designing an object or structure to satisfy physical constraints or minimize cost; working with typographic grid systems based on ratios).★

Number and Quantity: Quantities★

N-Q.1. Use units as a way to understand problems and to guide the solution of multi-step problems; choose and interpret units consistently in formulas; choose and interpret the scale and the origin in graphs and data displays.

N-Q.3. Choose a level of accuracy appropriate to limitations on measurement when reporting quantities.

Grade 8
Geometry

8.G.2. Understand that a two-dimensional figure is congruent to another if the second can be obtained from the first by a sequence of rotations, reflections, and translations; given two congruent figures, describe a sequence that exhibits the congruence between them.

Grade 7
Geometry

7.G.2. Draw (freehand, with ruler and protractor, and with technology) geometric shapes with given conditions. Focus on constructing triangles from three measures of

Project Putt-Putt

angles or sides, noticing when the conditions determine a unique triangle, more than one triangle, or no triangle.

7.G.6. Solve real-world and mathematical problems involving area, volume, and surface area of two- and three-dimensional objects composed of triangles, quadrilaterals, polygons, cubes, and right prisms.

Grade 6
Ratios and Proportional Relationships

6.RP.3d. Use ratio reasoning to convert measurement units; manipulate and transform units appropriately when multiplying or dividing quantities.

Answer Key
Before You Go: Uphill and Downhill

1. Slope

    $m = \Delta y/\Delta x$

    $m = 9/18$

    $m = 1/2$ or 50%

2. Slope

    $m = \Delta y/\Delta x$

    $m = 8/10$

    $m = 4/5$ or 80%

Check Yourself! Skill Check

1. Answers will vary.

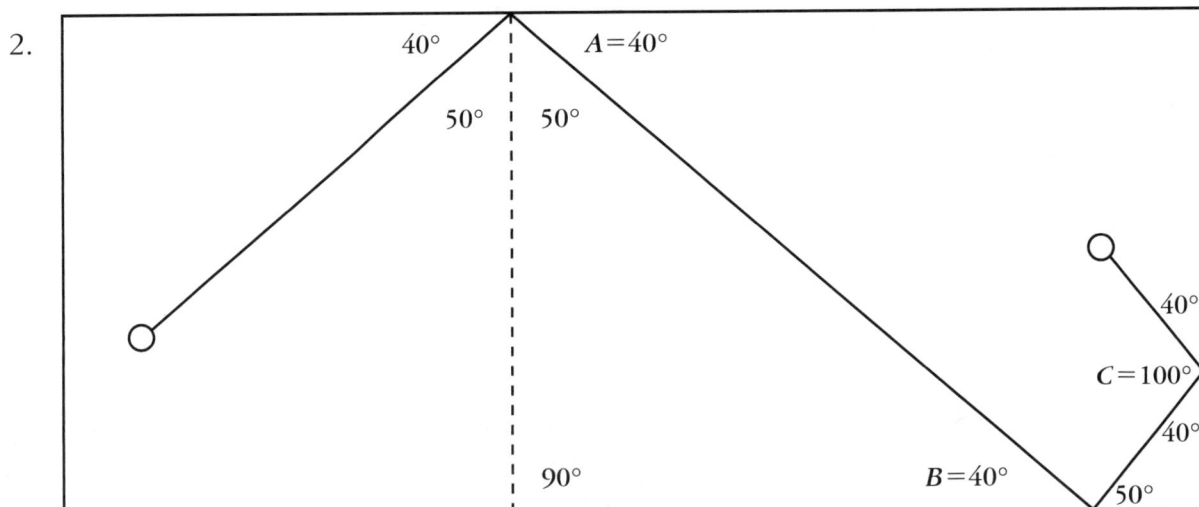

2.

Project Putt-Putt

Expedition Overview

Challenge
The International Pro Miniature Golf Tour is coming to your area. You are a tour champion turned course designer and have been asked to construct a unique and challenging course for the championship event. Tour planners want to see blueprints and a model immediately!

Objectives
- To explore transformational geometry and learn how to calculate angles of reflection
- To calculate slope and angle of slope
- To use measurement and geometry skills to create accurate scale drawings

Project Activities
Before You Go
- Reflection Inspection
- Uphill and Downhill

Off You Go
- Activity 1: Putt-Putt Blueprints
- Activity 2: Mini Model

Other Materials Needed
- golf balls
- graph paper
- paper
- colored pencils or crayons
- ruler
- protractor
- cardboard or poster board
- other recyclable construction materials (for example, paper towel rolls, containers, box covers, paper cups)
- other art supplies (for example, construction paper, felt, water-soluble paint, clay)
- scissors
- glue
- masking tape
- marbles
- wooden craft sticks

Project Putt-Putt

Expedition Overview

Lingo to Learn—Terms to Know

- angle of incidence
- angle of reflection
- area
- congruence
- isometry
- line of reflection
- perimeter
- ratio
- reflection
- scale drawing
- slope

Helpful Web Resources

- History of Miniature Golf
  www.terrastories.com/bearings/miniature-golf

- IgoUgo—A Field Guide to Mini-Golf
  www.igougo.com/story-s1214255-Myrtle_Beach-A_Field_Guide_to_Mini-Golf.html

- Professional Miniature Golf Association—Mini Golf Madness on The Travel Channel
  www.thepmga.com/Players/News/Mini_Golf_Madness/mini_golf_madness.php

- Professional Miniature Golf Association—Miniature Golf, Mathematically Speaking
  www.thepmga.com/Players/News/Geometry/geometry.php

- RekenWeb Games—KidsKount
  www.fi.uu.nl/rekenweb/en/welcome.xml
  (Click on "Mini Golf" link.)

- U.S. ProMiniGolf Association—U.S. Miniature Golf and Mini Golf Courses
  http://prominigolf.com/uscourses.html

Project Putt-Putt

Before You Go

Reflection Inspection

Goal:	To learn about reflection
Materials:	one golf ball per group, graph paper, masking tape, pencils

Directions

1. Form your group. Gather materials and put the golf ball aside.

2. Using graph paper, draw a line 4 to 5 inches long that represents a wall of your classroom. Decide how far from the line (wall) you want to position your hole. Mark this point on the graph paper. Label your hole point *H*. At the bottom of your graph paper, choose a point that represents your golf ball. Make the distance between this point and the line different from that of point *H* and the line. Label your ball point *B*.

3. Pause for group discussion. Make predictions about the path the ball will take for a hole in one. Take turns sharing how you would visualize the path and mentally gear up for the shot.

4. Draw the path on the grid. Label it "Shot 1."

5. Next, move the hole to a new location. Draw a second path. Label it "Shot 2."

6. Test your predictions. Select a section of wall. Use small pieces of masking tape to mark the ball starting point and your "hole."

7. Experiment with more shots by changing positions of the ball and the hole. If you have access to a space with two walls or solid vertical surfaces, try a few double-bounce or even triple-bounce shots. Don't forget to make predictions first.

8. Check your reflection prediction skills one more time. Show the path for a one-bounce or two-bounce hole in one below.

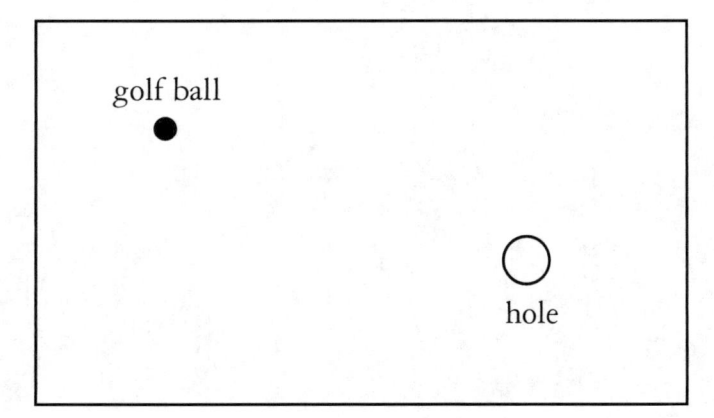

Expeditions in Your Classroom: Geometry © Walch Education

Project Putt-Putt

Before You Go

9. What other factors might affect the path of a golf ball on a miniature golf course? Write your ideas below.

Project Putt-Putt

Before You Go

Uphill and Downhill

> **Goal:** To review how to determine the slope and angle of slope of an incline

Miniature golf courses often include interesting topography, including sloped or graded surfaces such as ramps, steps, or embankments along the sides. How steep such an incline is can be described by calculating the slope.

Slope: $m = \dfrac{\Delta y}{\Delta x}$ or "rise over run" given as a ratio or percentage

Directions

Determine the slope for each diagram below.

1.

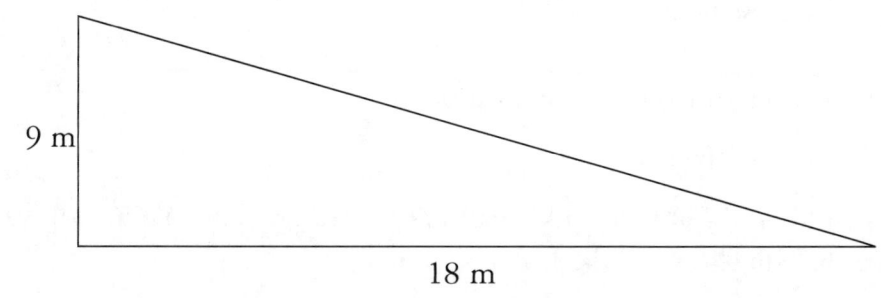

9 m

18 m

Slope: _____

2.

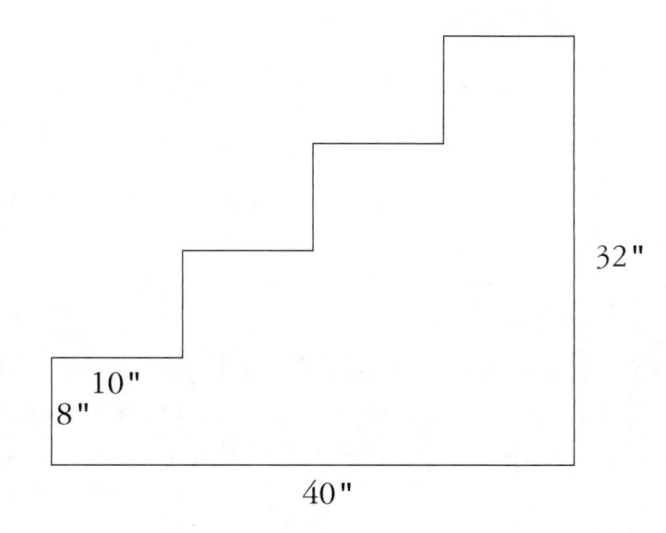

10"

8"

32"

40"

Slope: _____

Expeditions in Your Classroom: Geometry © Walch Education

Project Putt-Putt

Before You Go

Activity 1: Putt-Putt Blueprints

> **Goal:** To create a scale drawing for a miniature golf hole of your own design
>
> **Materials:** computer with Internet access, colored pencils or crayons, paper, graph paper, marbles

Directions

1. Use your Helpful Web Resources and the Internet to learn about miniature golf course design. Be sure to take a look at some of the world's most interesting and more well-known miniature golf courses.

2. Review the specifications for your miniature golf hole and develop three different hole ideas. Be creative! For example, choose a theme, a unique location, or interesting topography for your course hole.

> **Miniature Golf Course Hole Specifications**
>
> ❏ The "golf ball" will be a marble.
>
> ❏ Players should not be able to sink a ball in one straight shot. Your goal is to provide a challenge!
>
> ❏ The ball must bounce off course sides and/or obstacles at least twice for a hole in one.
>
> ❏ The hole must include obstacles, twists, or turns.
>
> ❏ The design should use at least three quadrants of graph paper.
>
> ❏ Include at least one incline or decline. Vertical drops may also be included.
>
> ❏ *Challenge:* There should be at least two unique shots a player could use to get a hole in one.

3. For each of your three ideas, create one freehand aerial view sketch. Show what your course hole would look like. Label important elements of the course (tee, hole, obstacles). Add other fun details—the name of the hole, a snack bar, a clubhouse, and so forth.

(continued)

Project Putt-Putt

Off You Go

4. Review your ideas. Get feedback. Test each concept! For example, use a small ball or marble and other temporary objects to test possible shots, angles, reflection, level of difficulty, and so forth.

5. Choose one idea. Create two accurate scale drawings for this idea: an aerial view and a cross-sectional (side) view. Follow the Scale Drawing Criteria below.

Miniature Golf Course Scale Drawing Criteria

- Draw the sketch accurately to scale.
- Provide a legend indicating scale.
- Use lines exactly where the ball would travel. *Tip:* Make a photocopy of your aerial view drawing and use the photocopy as a draft until you are sure you have the paths.
- Mark the points the ball must bounce off of for each hole in one.
- Provide distance and angle measurements. Include the following measurements:
  - perimeter of the course
  - height of sides
  - length and width of the course or course sections
  - location of the tee, hole, and other obstacles relative to sides and other course elements
  - total surface area of the hole
  - obstacle dimensions (height, width, clearance, etc. as appropriate)
  - diameter, circumference, and length of any tunnels included
  - slope and angle measurements for course inclines
  - height of any vertical drops
- Work is to be neat and colorful. Course elements should be well-labeled (tee, hole, obstacles).

General rule: If an item on your course can be measured, provide a measurement. For example, if you include a dinosaur obstacle, provide the height, the width at the base, the distance between the dinosaur's feet, and the clearance under the obstacle.

Project Putt-Putt

Off You Go

Activity 2: Mini Model

Goal:	To create a three-dimensional model of your miniature golf hole design
Materials:	cardboard, poster board, other recyclable materials (paper towel rolls, containers, cardboard boxes, box tops), construction paper, clay, other art supplies

Directions

1. Create a three-dimensional model of your course hole. Be sure to follow any additional instructions provided by your teacher.

Miniature Golf Course Hole Specifications

- Your model should be as large as your classroom desktop and no larger than two or three desktops. Follow any specific size instructions provided by your teacher.

- Your model must be constructed accurately to scale.

- Include all important course elements: course sides, tee, hole, incline/decline, and obstacles. You can also add creative touches or decorations that fit your theme. Use recyclable or reusable materials. For example, use construction paper for grass. A paper towel roll or a toilet paper roll can be cut and rolled more tightly to create a horizontal tunnel or vertical drop.

- Give your hole a name (for example, Dragon's Lair or Shipwreck Island).

- Mark points where the ball must bounce for a hole in one.

- Mark "par" for your hole in a visible location.

---

2. Bring your final model to class.

3. Set up holes around the classroom and play a few rounds of miniature golf!

Project Putt-Putt

Off You Go

Skill Check

1. You are in the final match of the National Pool Championship. Things haven't been going your way, but it's your turn now. You are solids. What is your shot strategy? Assume that no other balls move during shots except the cue ball. Because of your talent, the cue ball follows your shot and stops in good position for the next shot. You hit each of your balls straight on (no fancy spins). Number and illustrate the paths for your final series of shots. Mark the cue ball location for each shot with an X (X1, X2, X3, etc.).

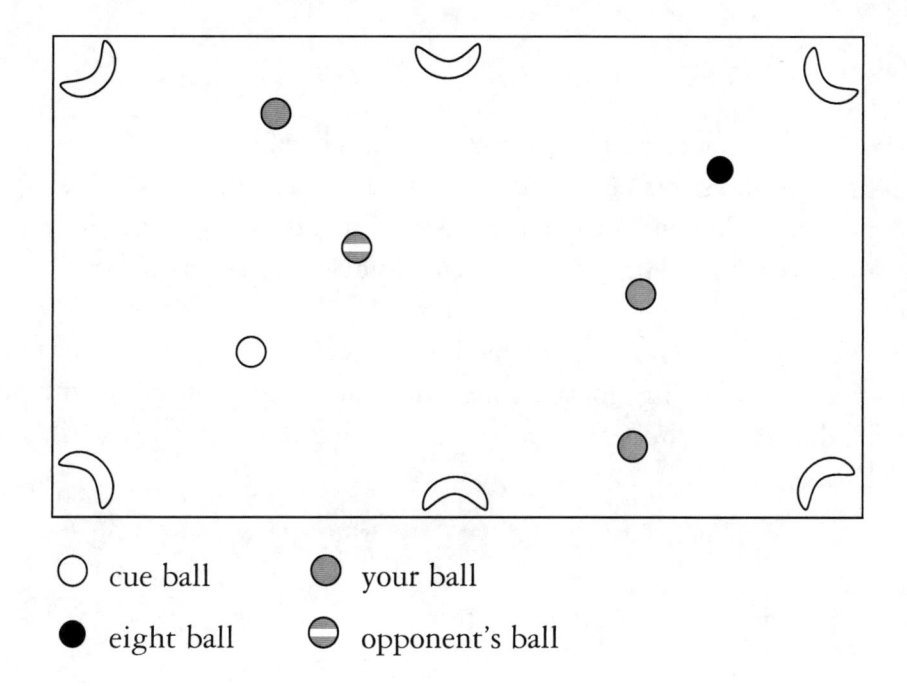

○ cue ball ● your ball
● eight ball ⊖ opponent's ball

2. What are the measures for angles A, B, and C in this highly skilled, three-bounce miniature golf shot?

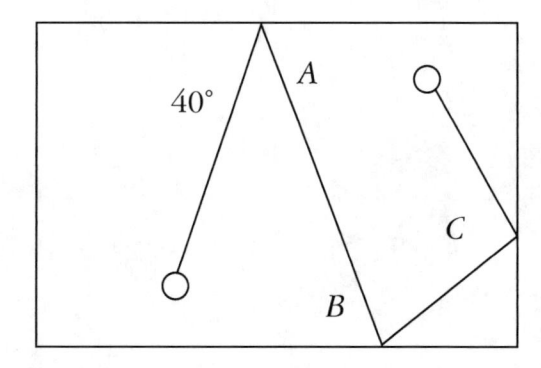

Expeditions in Your Classroom: Geometry

Project Putt-Putt

Check Yourself!

Self-Assessment and Reflection
Project Management

Before You Go

- ❏ I understand the concept of reflection. I know how to predict the angle a golf ball will reflect off a miniature golf course wall.
- ❏ I know how to calculate slope.
- ❏ I'm honestly not sure I understand the math involved and have asked my teacher for additional help.

Off You Go

- ❏ I reviewed our project challenge and project materials carefully.
- ❏ I understood the requirements of products I created: three concept sketches, two scale drawings for my best idea, and a three-dimensional model of my course hole.
- ❏ I carefully reviewed the Miniature Golf Course Hole Specifications before brainstorming ideas.
- ❏ I completed concept sketches for three hole ideas. I tested my ideas and got feedback.
- ❏ I completed two scale drawings for my course hole: an aerial view and a cross-sectional view. Both are drawn accurately to scale, include all required measurements, and are neat and well-labeled.

Do You Know?

- ❏ I can define the Lingo to Learn vocabulary terms for this project and give an example of each.
- ❏ I completed the Skill Check problems and carefully reviewed problems I answered incorrectly.

Project Putt-Putt

Check Yourself!

Reflection

1. What were the most challenging aspects of this project for you and why?

   _____
   _____
   _____
   _____

2. What skills did this project help you develop?

   _____
   _____
   _____
   _____

3. If you did this project again, what might you do differently and why?

   _____
   _____
   _____
   _____

Ripping Rooms

Overview
Students redesign the floor plan of a room. They provide an existing floor plan, concept drawings, and final plans for their new room.

Time
Total time: 8 to 10 hours plus one optional 2-hour activity

- Before You Go—Newspaper Room Design Challenge: one 55-minute class
- Before You Go—Miniaturize It: one 55-minute class
- Activity 1—My Old Room: 1 to 2 hours of homework
- Activity 2—My New Room Requirements: one 55-minute class (or as homework)
- Activity 3—Design Concept: one 55-minute class (or as homework)
- Activity 4—Final Room Drawings: 3 to 4 hours in class (or as homework)
- Activity 5—Materials Budget (Extra Challenge): 1 to 2 hours in class (or as homework)

Skill Focus
- polygon measurement
- two-dimensional models
- scale drawing

Prior Knowledge
- classification and attributes of polygons
- ratio and proportion
- Some exposure to floor plans is also helpful.

Team Formation
Students work individually, with partners, and in small groups of four to six students.

Lingo to Learn—Terms to Know
- **asymmetry:** lack of symmetry, balance, or proportion; imbalance
- **balance:** proportion; harmonious arrangement of parts within a whole
- **proportion:** an equation with one fraction set equal to another (two ratios are equivalent); balance among the parts of something
- **scale:** the ratio between the size of something and a representation of it
- **square footage:** the number of square feet; area calculated in feet

Expeditions in Your Classroom: Geometry

Ripping Rooms

- **surface area:** square units needed to cover a surface; sum of the areas of the faces of a three-dimensional shape
- **symmetry:** balance among the parts of something; exact reflection of a form on the opposite side of a dividing line or plane

Suggested Steps

Preparation

- Gather materials for each project.
- Find two or three examples of floor plans. (Use the Internet to search for key words "floor plan")
- If possible, find and print out pictures of common floor plan symbols.

Days 1 and 2

1. Provide an overview of the project and review materials.

2. Divide students into groups of four to six people for Before You Go: Newspaper Room Design Challenge. You need an even number of teams.

3. Support students by suggesting strategies for how the teams should work together during the design process.

4. Consider saving construction for the second day of the project.

5. Once construction is complete, prompt discussion. Ask the following questions:
   - Which structures were strongest? Weakest? Why?
   - Which shapes or structures were used and to what purpose?
   - Did drawings help or hinder construction? How difficult was it to draw the design?
   - How many teams made prototypes? How did this change their design? What effect did a prototype have on the final design?

Day 3

1. Review the concept of floor plans. Show samples to the class. Emphasize that floor plans use geometric shapes and symbols to represent features.

2. Introduce Before You Go: Miniaturize It.

3. Identify three or four objects that should appear in the class floor plan. Choose items that give students the measurement or calculation practice they need (circles, trapezoids, and so forth).

4. Assign individuals or pairs of students to collect measurements for the objects and the room.

Ripping Rooms

5. Have students work in groups of two to three people to create the floor plan.

6. Explain Activity 1: My Old Room. Show students the Current Room Plan Worksheet.

7. Indicate whether students will complete assignments for homework or if you will provide class time.

8. Give due dates for the Current Room Plan Worksheet and floor plan.

Homework

Have students complete the top half of the Current Room Plan Worksheet (room, window, door, and closet measurements). In addition to collecting room measurements, students may begin measuring furniture and other elements if desired.

Days 4 and 5

1. In class or for homework, have students sketch rough drafts of their room layouts using measurements they have.

2. Direct students to create lists of other major objects in their rooms. They should continue to collect measurements of the rooms and furnishings.

3. Check in with students regularly until floor plans are finished.

Day 6 (Current Room Plan Due)

1. Ask various students to showcase their room plans.

2. Talk about the activity. Ask students to describe their challenges. Have them explain their procedures.

3. Discuss if any students would wish to swap rooms (a partner designs your room and vice versa). Have students explain their thinking.

4. Introduce Activity 2: My New Room Requirements.

5. Divide students into pairs. Otherwise, students can work independently or can complete the Room Design Requirements list for homework.

6. Provide students with time to interview each other.

7. Ask students to prepare for the next stage of the project by looking for design ideas using the Internet, magazines, catalogs, and other resources.

Ripping Rooms

Homework

Have students complete the Room Design Requirements list. Also ask them to use web resources, magazines, and so forth to begin brainstorming design ideas.

Day 7

1. Explain Activity 3: Design Concept.

2. Indicate whether students will complete the tasks in class or for homework.

3. Remind students they should use geometric shapes, symbols, and labels to indicate floor plan elements.

4. Given the project goals, provide any other criteria you have for the new room. For example, you may want the design to include specific shapes students are studying; you may want to specify that the room shape cannot be square or rectangle; and so forth.

Homework

Have students review, revise, and complete their drawings.

Day 8

1. Invite a variety of students to show and explain their concept drawings.

2. Prompt discussion, highlighting the issues and information you expect to see in the final scale drawings. For example, if the room is square and all objects in it are square and positioned squarely, this may not be enough of a challenge for students. If items are positioned at an angle, you will want to see angle measurements in the final drawing. Emphasize any good floor-planning techniques you see (good use of shapes to represent objects and so forth).

3. Consider possible design award ideas to present on the last day. See Final Drawing Due Day on the following page.

4. Explain Activity 4: Final Room Drawings.

5. Review drawing criteria. Remind students that drawings should include furniture and other objects in the room.

6. Indicate if students should complete the final plans in class and/or for homework. Give a due date.

Tip: Students do not need to use one- or two-point drawing techniques for the side view, unless you specify otherwise. Indicate if students can use a computer to create all or part

Ripping Rooms

of their plans. If you have access to software that includes floor plan design features, use it. Designing in PowerPoint is possible, but less than ideal. Tell students they may also need to add other interior design features by hand, and that scale and proportion still count.

Final Drawing Due Day

1. Post final drawings around the classroom. Direct students to walk around to view work.

2. Have students select rooms for design awards. Give a variety of serious and silly awards, such as Most Ambitious Design, Most Geometric, Best Use of Protractor, Most Expensive Design, and Most Radical Change.

Final Day

1. Have students complete the Skill Check problems.

2. Check and review answers.

3. Have students complete the Self-Assessment and Reflection worksheet and submit it (optional).

Optional Materials Budget (Extra Challenge):

1. Explain the Materials Budget. Emphasize that you want students to get an estimated idea of their room cost. Students should not be expected to know the full extent of materials involved, although you might discuss examples. Students should choose two or three online stores to research prices.

2. Assign the budget for homework and give a due date (two or three days).

3. Invite students to share their findings. Discuss which rooms cost the most and least. Ask: Why? How difficult was it to calculate or estimate the amount of material needed?

Project Management Tips and Notes

- Students love scoping out potential features for their rooms. Let them start looking around from Day 1 of the project.
- For the final room drawings, students will have questions about the types of measurements you want them to use. Provide examples from concept drawings to illustrate the level of detail and elements you require. For example, you may have students focus on certain types of angle measurement, not on every angle in the room. Emphasize measurements that relate most directly to skills students need to practice.
- To simplify the final task, have students create only one drawing.

Ripping Rooms

Suggested Assessment
Use the Geometry Project Assessment Rubric or the following point system:

Team and class participation	10 points
Current room plan measurements	20 points
Current room plan scale drawing	20 points
Two new plan scale drawings	50 points
Materials budget	Bonus points

Extension Activities
- Explain the one-point drawing technique (perspective drawing). Have students create the cross-sectional drawing of the new room using the technique.
- Invite an architect, a contractor, or an interior designer to class to show students how to read blueprints, floor plans, and interior design plans.

Other Helpful Resources
Suite 101—Understanding One-Point Perspective
http://alinabradford.suite101.com/understanding-onepoint-perspective-a67256

Common Core State Standards Connection
High School
Geometry: Modeling with Geometry

G-MG.1. Use geometric shapes, their measures, and their properties to describe objects (e.g., modeling a tree trunk or a human torso as a cylinder).★

G-MG.3. Apply geometric methods to solve design problems (e.g., designing an object or structure to satisfy physical constraints or minimize cost; working with typographic grid systems based on ratios).★

Number and Quantity: Quantities★

N-Q.1. Use units as a way to understand problems and to guide the solution of multi-step problems; choose and interpret units consistently in formulas; choose and interpret the scale and the origin in graphs and data displays.

Grade 7
Geometry

7.G.1. Solve problems involving scale drawings of geometric figures, including computing actual lengths and areas from a scale drawing and reproducing a scale drawing at a different scale.

Expeditions in Your Classroom: Geometry

Ripping Rooms

7.G.6. Solve real-world and mathematical problems involving area, volume, and surface area of two- and three-dimensional objects composed of triangles, quadrilaterals, polygons, cubes, and right prisms.

Grade 6
Geometry

6.G.1. Find the area of right triangles, other triangles, special quadrilaterals, and polygons by composing into rectangles or decomposing into triangles and other shapes; apply these techniques in the context of solving real-world and mathematical problems.

Ratios and Proportional Relationships

6.RP.3d. Use ratio reasoning to convert measurement units; manipulate and transform units appropriately when multiplying or dividing quantities.

Answer Key
Check Yourself! Skill Check

1.

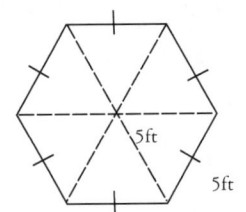

   Calculate the area of one equilateral triangle (two right triangles) and multiply by 6.
   Sides of equilateral triangle:
   $a^2 + b^2 = c^2$
   $2.5^2 + b^2 = 5^2$
   $b = 4.3$ feet
   Area of equilateral triangle:
   $A = \frac{1}{2}bh$
   $A = \frac{1}{2}(5 \times 4.3)$
   $A = 10.75$
   Area of the hot tub:
   $6 \times 10.75 = 64.5$ square feet

2. A = area of the rectangle—area of the missing corner/triangle
   $A = (bh) - (\frac{1}{2}bh)$
   $A = (8 \times 6) - \frac{1}{2}(3 \times 3)$
   $A = 48 - 4.5$
   $A = 43.5$ square feet

3. a. 150°
   b. 150°

Ripping Rooms

Expedition Overview

Challenge
You have just won a national contest sponsored by *Ripping Rooms,* a popular home improvement magazine. Your prize? *Ripping Rooms* will pay to have your bedroom redesigned to your specifications. They just need to know what you want to do. They want the design ideas to come from you. After all, it is your space, and you have to live in it.

Ripping Rooms needs you to provide the following: a floor plan of your existing room, an outline of room redesign specifications, two scaled drawings that show your redesign vision, and an estimate of costs. For this project, you can handle the redesign yourself or turn it over to a fresh pair of eyes (another student in the class).

Objectives
- To create and interpret scale drawings
- To demonstrate understanding of ratio and proportion
- To calculate the dimensions of geometric figures
- To apply geometry to architecture, construction, and interior design

Project Activities
Before You Go
- Newspaper Room Design Challenge
- Miniaturize It

Off You Go
- Activity 1: My Old Room
- Activity 2: My New Room Requirements
- Activity 3: Design Concept
- Activity 4: Final Room Drawings
- Activity 5: Materials Budget (Extra Challenge)

Expedition Tool
Current Room Plan Worksheet

Other Materials Needed
- newspapers
- masking tape
- paper
- graph paper
- colored pencils or crayons
- ruler
- protractor
- tape measure or roll of string
- digital camera (optional)

Ripping Rooms

Expedition Overview

Lingo to Learn—Terms to Know
- asymmetry
- balance
- proportion
- scale
- square footage
- surface area
- symmetry

Helpful Web Resources
- Microsoft Office Visio—Create a Floor Plan
  http://office.microsoft.com/en-us/visio-help/create-a-floor-plan-HP001208559.aspx
- The Museum of Modern Art—Red Studio: You Design
  http://redstudio.moma.org/interactive
- Suite 101—Understanding One-Point Perspective
  http://alinabradford.suite101.com/understanding-onepoint-perspective-a67256

Ripping Rooms

Before You Go

Newspaper Room Design Challenge

Goal:	To think about design, planning, structure, and construction
Materials:	10 sheets of newspaper per team, 10 inches of masking tape per team member

Directions

1. Form a team.

2. As a team, create a plan, an illustration, and directions for building a free-standing structure. You will trade plans with another team and construct each other's design. Your plan must consider the following criteria:

   Newspaper Structure Criteria

   ❑ Builders can only use 10 sheets of newspaper per team and 10 inches of masking tape per team member.

   ❑ The structure must allow team members to enter and exit the structure without changing it to do so.

   ❑ All team members must be able to stand inside the structure at the same time.

3. Be careful about how you use materials. You may use some of your materials to construct a prototype; however, you still have to provide the other team with enough materials to construct the design. If you create a prototype, make sure materials can be returned to their original state and reused by the other team.

4. Once your instructions are written, trade your plans with the other team.

5. Using the drawing, instructions, and materials provided, build the structure designed by the other team. You may only speak with members of your own team.

6. Once the structures are built, share and discuss the experience.

Ripping Rooms

Before You Go

Miniaturize It

> **Goal:** To visualize floor plans and scale two- and three-dimensional objects
>
> **Materials:** two or three tape measures, paper, graph paper, pencil, rulers, protractor

Directions

1. Imagine your classroom as seen from above. In the space below, sketch the layout and objects in the room.

2. Get your measurement assignment from your teacher.

3. Record your measurements. Report them to the class.

4. Write down all measurement information provided by classmates.

5. On graph paper, create a scale floor plan of your classroom based on the measurements reported. Put the objects in the correct location, using the correct size and proportion.

6. Review the results. Compare the similarities and differences among the drawings. Discuss the challenges of scale drawing.

Ripping Rooms

Off You Go

Activity 1: My Old Room

> **Goal:** To measure and create a floor plan for your existing bedroom
>
> **Tools:** Current Room Plan Worksheet

Ripping Rooms needs to know the basic structure of your bedroom. They want to feature "before" and "after" views of your room in the final article.

Directions

1. Use the Current Room Plan Worksheet to record measurements of your current bedroom. Include dimensions of windows, doors, closets, pieces of furniture, and other items.

2. Draw or use your computer to create a floor plan of your room.

> **Current Room Floor Plan Criteria**
>
> ❑ Your floor plan should be two-dimensional and drawn to scale. For example, if you use $1/4$-inch graph paper, each square might equal $1/4$ foot.
>
> ❑ Represent furniture and other objects to scale.
>
> ❑ Use basic shapes and symbols to represent objects and features. For example, a rectangle might represent a bed, a circle might represent a lamp, and a polygon might represent a computer.
>
> ❑ Label everything. Provide a legend that indicates scale.
>
> ❑ Note dimensions for all items. Your floor plan will be two-dimensional, so you won't need to include height measurements.

3. *Optional:* Take pictures of your room and objects in it, especially anything you want to redesign.

Ripping Rooms

Expedition Tool

Current Room Plan Worksheet

Use this worksheet to record measurements of your current bedroom and its contents.

Length of room: _____

Width of room: _____

Height of room: _____

Total square footage: _____

Total surface area (walls): _____

Other dimensions (if your room is an irregular shape): _____
_____

Window	Height	Width
1		
2		
3		

Door	Height	Width
1		
2		

Closet	Height	Width	Depth
1			
2			

Ripping Rooms

Expedition Tool

In the space below, list the objects in your room. Provide basic dimensions: length, width, height, perimeter/circumference, and surface area.

Item	Basic Shape (aerial view)	Dimensions

Expeditions in Your Classroom: Geometry © Walch Education

Ripping Rooms

Off You Go

Activity 2: My New Room Requirements

Goal: To outline the requirements of your new bedroom

Directions

1. Be a do-it-yourselfer or trade rooms with a classmate. If swapping, trade floor plans and show any photos you have to your partner.

2. Brainstorm and outline the features and requirements of your redesigned room. Name the activities you should be able to do in the room. For example, maybe you want to swim in your own indoor swimming pool or view the stars. Record requirements on the lines below.

Room Design Requirements

Examples:

Must have extra-large shoe racks for size-13 feet.

Must have soundproof booth for tuba practice.

1. _____
2. _____
3. _____
4. _____
5. _____
6. _____
7. _____
8. _____
9. _____
10. _____

Ripping Rooms

Off You Go

Activity 3: Design Concept

Goal:	To sketch your room redesign ideas
Materials:	paper or graph paper, pencil, colored pencils or crayons, computer with Internet access, magazines or catalogs

Directions

1. Brainstorm ideas for the new room. Sketch your ideas on another sheet of paper or graph paper. Get inspiration from Internet sites, or explore furniture and home-improvement stores. Browse magazines or catalogs.

   Note: Bookmark Internet sites so that you can find them again easily. You will need to gather prices to submit a cost estimate for the project. You will also want to check furniture dimensions later.

2. Create two simple freehand sketches that depict your proposed design changes. One drawing should show the floor plan of the new room. The other should show a side view. These should be simple "concept" drawings that show only the basic layout of the room and any structural changes that affect the floor plan, major features, and the placement of important objects. Use basic shapes or symbols to represent objects. If you are more artistic, draw in the details. You can also print or copy images from web sites or magazines. Whatever your method, try to represent the scale and proportion of items as accurately as possible.

3. Find a partner to work with and get feedback on your design. Use these guiding questions:

   - Does the design meet the requirements outlined earlier? If not, why not?
   - Is the design logical and user-friendly? For example, what are the most common paths one would take to accomplish primary tasks? Are high-traffic areas free and clear?
   - Is there balance to the design? Was the symmetrical or asymmetrical approach successful?
   - What is the relationship between the inhabitant, the objects, and the room? Is there good proportion?
   - Is it safe? Where is the emergency exit? What might go wrong and how might it be fixed?

Expeditions in Your Classroom: Geometry

Ripping Rooms

Off You Go

Activity 4: Final Room Drawings

> **Goal:** To create two scaled drawings that show the final redesign
>
> **Materials:** paper or graph paper, pencil, colored pencils or crayons, ruler, protractor, computer (optional)

Directions

1. Use any feedback you received to modify your design ideas and sketches.

2. Create two scale drawings for the redesigned room: a floor plan and a cross-sectional or side view. Choose the perspective that provides the best view.

> **Scale Room Drawings Criteria**
>
> ❏ Both drawings should be neat and legible.
>
> ❏ Represent all room features in view—windows, doors, closets, basketball hoops, beanbag chairs, and so forth. Use symbols and geometric shapes to signify items.
>
> ❏ Label dimensions for all features and items with true or realistic dimensions. For example, a mattress that you label 4' × 5' is too short for an actual bed. If needed, use the Internet to find standard item sizes or measure true objects.
>
> ❏ Define angle measurements according to your teacher's instructions.
>
> ❏ Provide a legend to indicate scale.
>
> ❏ Label all items shown.
>
> ❏ Show electrical features (lights, outlets, switches, cable connection, and so forth).

Note: Remember, scale drawings take effort—and erasing! Create a rough draft you can use to position items, work out measurements, and adjust. Use a fresh sheet of paper for your final copy.

3. Optional: Use a computer to create your final drawings.

Ripping Rooms

Off You Go

Activity 5: Materials Budget (Challenge)

> **Goal:** To provide a list of estimated materials and costs for your room redesign project
>
> **Materials:** computer with Internet access, a variety of home-repair and home-decor catalogs

Provide a list of materials and estimated costs for your project. Use the Material Budget Worksheet to guide you. Create your final estimate using your computer and spreadsheet software.

Directions

1. List the materials you need. Include building materials, materials for interior design, furnishings, and equipment. Make your best guess about what you need.

2. Use the Internet or catalogs to research item costs. Provide prices for items you know or can find easily—for instance, the cost of a bed frame, a rug, or a hot tub. Estimate other costs.

3. Research materials such as flooring, tile, paint, or fabric. Calculate the total surface area you need to cover. Find out how much one unit of the item costs (1 gallon of paint, 1 yard of fabric, or 1 sheet of drywall) and estimate the total amount you need. If you cannot find the exact item you need for your room design, use something similar to make your calculations.

4. Estimate materials such as the wood needed to build a new wall. Make assumptions! For example: For this wall, I will need 1 two-by-four spaced every 12 inches.

Ripping Rooms
Off You Go

Material Budget Worksheet

Item	Assumptions (how amount is determined)	Cost per unit	Number of units needed	Subtotal
		Total project cost (excluding labor)	

Ripping Rooms

Check Yourself!

Skill Check

1. The hexagon below represents a hot tub. How much space will the hot tub occupy?

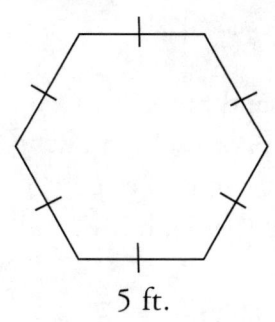

   5 ft.

2. What is the area of the room illustrated below?

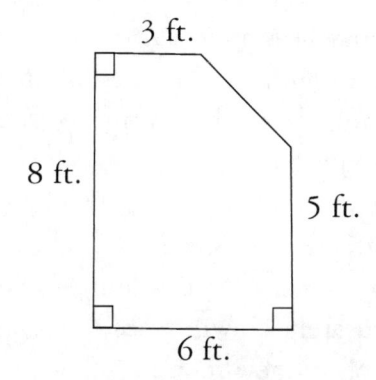

   3 ft.
   8 ft.
   5 ft.
   6 ft.

3. The figure below represents a new deck you want to build for your house.

   a. What is the measure of angle X?

   b. What is the measure of angle Y?

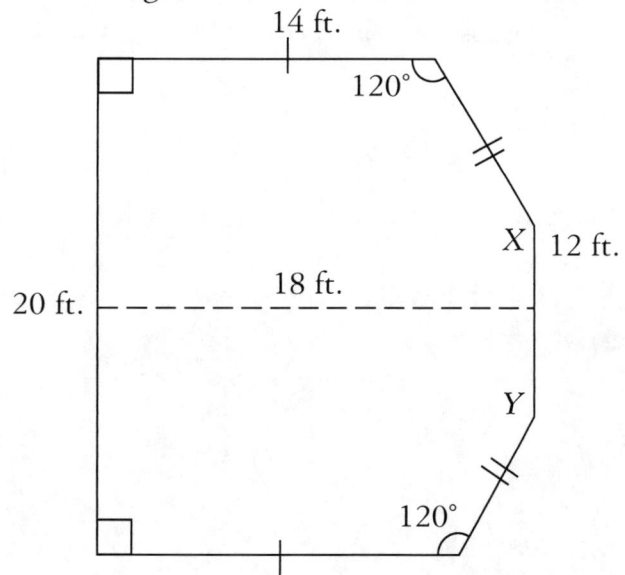

   14 ft.
   120°
   X 12 ft.
   20 ft.   18 ft.
   Y
   120°

Expeditions in Your Classroom: Geometry

Ripping Rooms

Check Yourself!

Self-Assessment and Reflection
Project Management

Before You Go

- ❑ I understand the concepts of ratio and proportion.
- ❑ I understand what it means to prepare a scale drawing or floor plan.
- ❑ I know how to calculate surface area and other polygon measurements.
- ❑ I'm honestly not sure I understand the math involved and have asked my teacher for additional help.

Off You Go

- ❑ I reviewed our project challenge and project materials carefully and thoroughly. I understood the requirements of products I needed to create: a scaled floor plan of my current bedroom, two preliminary sketches of my redesigned room, and two final scaled drawings of the new room plan.
- ❑ I carefully reviewed drawing specifications. I received clarification on any questions about dimensions and measurements I needed to include.
- ❑ I took as many measurements in my current room as I could.
- ❑ My current room plan is accurately drawn to scale and well-labeled.
- ❑ I outlined the requirements of my new room.
- ❑ I completed two freehand sketches of my design ideas—a floor plan view and a side view. I got feedback on my design.
- ❑ I created two accurately scaled plans of the new room—a floor plan and a side view. My plans are detailed, have clear labels, and include dimensions.
- ❑ *Optional:* I researched the cost of items involved in my room redesign and created a list of material expenses.

Do You Know?

- ❑ I can define the Lingo to Learn vocabulary terms for this project and give an example of each.
- ❑ I completed the Skill Check problems and carefully reviewed problems I answered incorrectly.

Ripping Rooms

Check Yourself!

Reflection

1. What were the most challenging aspects of this project for you and why?

   _____
   _____
   _____
   _____
   _____

2. What skills did this project help you develop?

   _____
   _____
   _____
   _____
   _____

3. If you did this project again, what might you do differently and why?

   _____
   _____
   _____
   _____
   _____
   _____

Fashionistas

Overview
Students investigate geometry in fashion design and clothing construction, and then create patterns for actual and newly imagined articles of clothing. Then they "sew" paper shirts and model them.

Time
Total time: 8 to 10 hours

- Before You Go—Figure Figures: one 55-minute class period
- Activity 1—Fashion Geometry: one 55-minute class and 45 to 60 minutes of homework
- Activity 2—Clothes Inspection: one to two 55-minute class periods and 10 minutes of homework
- Activity 3—You Be the Designer: two 55-minute class periods and 30 to 60 minutes of homework
- Activity 4—Paper Clothes Fashion Show: one to two 55-minute class periods

Skill Focus
- measurement
- ratio and proportion
- visualization and spatial reasoning
- representation and construction of three-dimensional objects

Prior Knowledge
- measurement skills
- basic understanding of ratio and proportion

Team Formation
Students can work individually or in teams of two or more for specified activities. See Project Management Tips and Notes.

Lingo to Learn—Terms to Know
- **proportion:** an equation with one fraction set equal to another (two ratios are equivalent); balance among the parts of something
- **ratio:** a pair of numbers that compares different types of units
- **symmetry:** balance among the parts of something; exact reflection of a form on the opposite side of a dividing line or plane

Expeditions in Your Classroom: Geometry

Fashionistas

Suggested Steps
Preparation

- Review the list of materials and collect anything you will provide.
- Borrow or buy a sewing pattern to help introduce the project.
- Find pictures of four or five examples of "fashion geometry" in advance. Examples should cross time, gender, and style. Examples: armor, kilts, football jerseys, hoop skirts, top hats, mod styles from the 1960s, Elizabethan doublets, protective clothing for firefighters or military personnel, and so forth.
- Gather one tape measure per student to be used for three to four class periods. Be sure to decide on metric or standard as the unit.
- Some students may not be comfortable creating a pattern based on their actual body measurements. Prepare a set of generic measurements to use as a substitute. Find a willing volunteer (a teacher), or use the list of body measurements on the Clothes Inspection Worksheet as your guide.
- Assemble a collection of decorative supplies students can use on their final shirts (feathers, glitter, and so forth).

Day 1

1. Provide an overview of the project and review project materials.
2. Solicit a few examples of geometry in clothing students are wearing that day. Share examples from history such as armor, kilts, football jerseys, hoop skirts, top hats, and so forth.
3. Explain that students will create a sewing pattern during the project. Show a sample pattern. Ask students what they notice about it. Discuss how a pattern is similar to a house blueprint.
4. Facilitate Before You Go: Figure Figures to help students understand basic body proportion.

Homework

Have students complete Activity 1: Fashion Geometry and prepare pictures or sketches of five examples to share in class.

Fashionistas

Day 2

1. Review the examples students found for Activity 1: Fashion Geometry. Add more of your own to help students see the range.

2. Invite students to suggest why particular geometric patterns, shapes, cuts, or styles were used. Talk about function, features, and style.

3. Preview Activity 2: Clothes Inspection. Assign the first step—finding an article of clothing—for homework.

4. Ask students to bring in a tape measure if possible.

Homework

Have students find an article of clothing to bring to class.

Day 3

1. Ask students to examine the article of clothing closely.

2. Distribute tape measures. Explain the Clothes Inspection Worksheet. Indicate whether students should use metric or standard measurements.

   Note: The more complex the clothing, the more challenging this will be! Support students by explaining that they should visualize the main pieces. You may even give them permission to disregard linings and other interior features (for example, for a coat or jacket).

3. Some pieces might include curves or tricky shapes. Ask the class for suggestions on how to get a measurement or fairly accurate estimate of an odd shape. Work on one example together.

4. Have students sketch the shapes of the pieces they think were used to construct their article of clothing, and measure or calculate dimensions. Students might use different approaches. Some may sketch first and then find measurements. Others might want measurements to help them draw the piece. Either approach is fine.

5. Leave 10 minutes for discussion (challenges, observations). Students may only have sketches and measurements for three or four pattern pieces; this is fine. If you think students need more measurement practice, have them finish up for homework. Otherwise, end the activity as is.

Fashionistas

Day 4

1. Review Activity 3: You Be the Designer. Note that this stage of the project will take several class periods and homework assignments.

2. Let students sit in pairs or groups to brainstorm design ideas.

3. Direct students to work independently or in pairs to sketch their final idea on the My Design worksheet.

4. Ask students to take body measurements for their pattern. They can use their own measurements, those of a classmate, or the standard set you provide. Students who used a shirt for Activity 1: Fashion Geometry may already have some or all of the measurements they need.

5. Explain the idea of "ease"—adding $\frac{1}{2}$ to 1 inch to certain measurements for a looser fit (for example, under the arms).

6. If time allows, have students start sketching pattern pieces. These do not need to be drawn precisely to scale. You want students to be clear about the pieces first. Ask students to complete this step for homework.

Homework

Have students complete rough sketches of pattern pieces. If they wish, students can begin the next step—figuring out piece dimensions.

Day 5

Note: Pattern creation should be done in class and may require more than one period.

1. Review the next step of the pattern construction process: figuring out other measurements needed to create pattern pieces. This will take a combination of strategies (taking more body measurements, draping paper to test fit, and performing calculations). Encourage students to help one another. If you notice someone using a good strategy, highlight it for the class. Show sample sewing pattern pieces again. Remind students that their pieces won't look exactly like a completed professional pattern.

2. Sign off on each student's pattern sketches and measurements. Provide the butcher paper to students who are ready to move on to drawing. They should not cut pieces yet.

Fashionistas

3. Remind students of their three tasks: drawing the pattern pieces according to the original measurements; adding a cutting line; and adding a line to show how to size the pieces for someone 8% larger or smaller (they choose which).

4. Review pattern drawings with each student. Make sure each piece is numbered, labeled, and includes three lines (original size, cutting line, second size). If complete, let the student cut pieces out.

Day 6

1. Continue to support students as they create the patterns. Encourage students who finish early to help other students; however, direct them to do tasks (cutting, connecting lines), not "thinking" work (plotting points for lines, scaling).

2. Have all students finish pattern creation before continuing to the final step of the project.

3. When pattern pieces are cut and ready, give students their final batch of butcher paper.

4. Instruct students to trace their pattern onto the paper, cut out pieces, and assemble them using tape. Remind students to use tape sparingly in strategic locations to hold things together during assembly and fitting until they are sure about final assembly.

5. Show students the art supplies they can use to decorate their clothing. Avoid paint and other supplies that need to dry overnight.

6. Ask students to wear and model their creations. If you have time, have fun with this—choose students to serve as judges (rotate the role), set up a runway area, and so forth.

7. Have students display their designs around the classroom.

8. Debrief the activity. See Activity 4: Paper Clothes Fashion Show for discussion questions.

Final Day

1. Have students complete the Skill Check problems.

2. Check and review answers.

3. Have students complete the Self-Assessment and Reflection worksheet and submit it (optional).

Fashionistas

Project Management Tips and Notes
- As written, the project specifies that students design a shirt, pants, or a dress. You may have everyone work on the same item or give more options.
- To simplify Activity 2, ask students to bring in the same type of clothing (for example, a T-shirt).
- To simplify Activity 3, have all students design a shirt pattern for the same size person, using body measurements you provide. Have them work in teams to create the pattern. Allow slight variations in their T-shirt design (long sleeves, V-necks) or not. Ask students to size the pattern up (no choice).
- To simplify Activity 4, have students work in teams to create and decorate one shirt. This option takes away individual variety but allows for more teamwork.

Suggested Assessment
Use the Geometry Project Assessment Rubric or the following point system:

Team and class participation	10 points
Clothing inspection	15 points
Clothing design sketch	20 points
Pattern	50 points
Project self-assessment	5 points

Extension Activities
- Technologically inclined students may use computer software to design patterns.
- Have students develop a shirt for a charitable cause or event. Students create the pattern, design the item, and construct it.
- Have students learn more about "body geometry"—how clothing is designed to accommodate body shape and motion, or how clothing is engineered for special occupations (such as for astronauts, military personnel, or athletes).

Common Core State Standards Connection
High School
Geometry: Modeling with Geometry

G-MG.1. Use geometric shapes, their measures, and their properties to describe objects (e.g., modeling a tree trunk or a human torso as a cylinder).★

G-MG.3. Apply geometric methods to solve design problems (e.g., designing an object or structure to satisfy physical constraints or minimize cost; working with typographic grid systems based on ratios).★

Fashionistas

Number and Quantity: Quantities★

N-Q.1. Use units as a way to understand problems and to guide the solution of multi-step problems; choose and interpret units consistently in formulas; choose and interpret the scale and the origin in graphs and data displays.

N-Q.3. Choose a level of accuracy appropriate to limitations on measurement when reporting quantities.

Grade 7
Geometry

7.G.1. Solve problems involving scale drawings of geometric figures, including computing actual lengths and areas from a scale drawing and reproducing a scale drawing at a different scale.

7.G.6. Solve real-world and mathematical problems involving area, volume, and surface area of two- and three-dimensional objects composed of triangles, quadrilaterals, polygons, cubes, and right prisms.

Ratios and Proportional Relationships

7.RP.2a. Decide whether two quantities are in a proportional relationship, e.g., by testing for equivalent ratios in a table or graphing on a coordinate plane and observing whether the graph is a straight line through the origin.

Grade 6
Ratios and Proportional Relationships

6.RP.1. Understand the concept of a ratio and use ratio language to describe a ratio relationship between two quantities. *For example, "The ratio of wings to beaks in the bird house at the zoo was 2:1, because for every 2 wings there was 1 beak." "For every vote candidate A received, candidate C received nearly three votes."*

6.RP.3d. Use ratio reasoning to convert measurement units; manipulate and transform units appropriately when multiplying or dividing quantities.

Answer Key
Before You Go: Figure Figures

1. Every size increases each measurement by 2 inches. Each size is related to the previous size in the same way. Size 10 is approximately 6% bigger in the chest than size 8 and 8% bigger in the waist. Size 18 is approximately 5% bigger in the chest than size 16 and 6% bigger in the waist. The proportion from size to size changes very slightly.

2. Chest and waist differ by the same amount, 8 inches, at each size. They are proportionate. Chest is approximately 25% bigger than waist.

Fashionistas

3. Each of these areas represents $1/3$ of the croquis. This is the typical proportion for a croquis drawing. You might ask students to check their proportions to compare.
4. They are evenly spread, or proportionate.
5. An arm is typically 25% to 33% longer. Some croquis elongate the proportion to make the model seem more glamorous. You might ask students to check their own arm-to-body proportion.

Activity 4: Paper Clothes Fashion Show
6–11. Answers will vary.

Check Yourself! Skill Check
1.

	Size 12
Chest	35 inches
Waist	28 inches
Hips	37 inches

Compared to size 8, size 10 is 2" bigger in the chest, 3" bigger in the waist, and 4" bigger in the hips. Increase size 10 by these amounts to create size 12. Round up when needed.

2. a. She could make the dress.
   b. She has 6 feet × 3 feet = 18 square feet (2,592 square inches) of fabric.

   Pants
   Rectangular area pieces cut from: 42 inches × 36 inches = 1,512 square inches × 2 pieces (front and back) = 3,024 square inches
   Cutting two pieces requires $2\frac{1}{3}$ yards of fabric. Pants must be cut vertically.
   Area of pants pieces = 1,512 square inches – 532 square inches (scraps left after cutting) = 980 square inches × 2 pieces = 1,960 square inches

   Dress
   Rectangular area pieces cut from: 36 inches × 25 inches = 900 square inches × 2 pieces (front and back) = 1,800 square inches
   Cutting two pieces requires 50 inches or 1.39 yards. Dress pieces can be cut horizontally.
   Actual area of dress = 900 square inches – 169.125 square inches (scraps left after cutting) = 730.875 square inches × 2 pieces = $1461\frac{3}{4}$ square inches

   b. Pants: She would need $1/3$ yard (1 foot).
Dress: She will have 0.61 yards of uncut fabric. She also has four irregular polygon scrap pieces, each 3 yards, with an area of 84.5625 square inches. This is great for a few quilt squares but not enough for additional clothes.

Fashionistas

Expedition Overview

Challenge
Have you been bitten by the fashion bug? Or are you the type to throw on anything you find in the clean laundry pile without thinking twice? Whether you go glam or go grunge, your clothes are the end result of a dynamic mix of art, design, and engineering. This project lets you explore that intersection and test your own creative and mathematical expertise.

Objectives
- To use visualization and spatial-reasoning skills
- To create patterns or blueprints for three-dimensional objects
- To practice measurement, ratio, and proportion

Project Activities
Before You Go
- Figure Figures

Off You Go
- Activity 1: Fashion Geometry
- Activity 2: Clothes Inspection
- Activity 3: You Be the Designer
- Activity 4: Paper Clothes Fashion Show

Expedition Tools
- Croquis Worksheet
- Clothes Inspection Worksheet
- My Design

Other Materials Needed
- tape measures (one per student)
- pencils
- colored pencils
- graph paper
- one to two rolls of butcher paper
- transparent tape
- scissors (one pair per student)
- art supplies (markers, construction paper, glitter, glue, feathers, and so forth)

Fashionistas

Expedition Overview

Lingo to Learn—Terms to Know
- proportion
- ratio
- symmetry

Helpful Web Resources
- Bravo—Project Runway
  www.bravotv.com/Project_Runway

- BurdaStyle—Figurine for Technical Drawing
  www.burdastyle.com/techniques/figurine-for-technical-drawing

- Butterick—Butterick History
  www.butterick.com/bhc/pages/articles/histpgs/about.html

- College Board—Career: Fashion Designers
  www.collegeboard.com/csearch/majors_careers/profiles/careers/105101.html

- Fashion Museum
  www.fashionmuseum.co.uk

- Fashion: Past & Present
  www.teacheroz.com/fashionhistory.htm

- Home Sewing Association
  www.sewing.org

- Home Sewing Association—Teen Sewing & Craft Projects
  www.sewing.org/html/teen.php

- PBS: What Is Fashion?
  www.pbs.org/newshour/infocus/fashion/whatisfashion.html

- Taunton's Threads—Meet the *Threads* Croquis Family: Your Tool for Fashion Sketching
  www.taunton.com/threads/pages/t00147.asp

- *Threads* Magazine—How to Measure Up
  www.threadsmagazine.com/item/4610/how-to-measure-up

Fashionistas

Before You Go

Figure Figures

Goal:	To practice ratio and proportion, and learn about fashion sketching
Materials:	pencil; ruler or tape measure
Tools:	Croquis Worksheet

Directions

The chart below is a clothes sizing chart for women's wet suits. Use it to answer the questions that follow.

Size	8	10	12	14	16	18	20
Chest (inches)	30–32	32–34	34–36	36–38	38–40	40–42	42–44
Waist (inches)	22–24	24–26	26–28	28–30	30–32	32–34	34–36

1. What is the relationship between sizes 8 and 10? Are they proportionate? Are sizes 16 and 18 related in the same way?

   _____
   _____
   _____

2. What is the relationship between chest and waist measurements?

   _____
   _____
   _____

In fashion design, a *croquis* is a quick sketch of a model or a template that illustrates human body proportions. It is used to show how clothing will look. Typically, a croquis is drawn "9 heads" tall, the accepted height proportion for fashion illustration. Men and women share the same height proportion. A designer tries to retain the basic proportions of the head and torso but might lengthen the legs to make a design look more elegant.

(continued)

Expeditions in Your Classroom: Geometry © Walch Education

Fashionistas

Before You Go

Look at the croquis on the following page. Choose one, take measurements, and explore proportion in the model. Answer the following based on the model you chose.

3. Describe the relationship between the following three areas:
   - top of head to waist
   - waist to knees
   - knees to bottom of feet

   _____
   _____
   _____
   _____

4. What is the relationship between the lines you drew in the upper portion of the body (shoulder, chest, waist, end of torso)?

   _____
   _____
   _____
   _____

5. What is the relationship between arm length and shoulder-to-waist measurement?

   _____
   _____
   _____
   _____

Fashionistas

Expedition Tool

Croquis Worksheet

Choose a croquis below. Draw horizontal lines across the model at the following points. Extend your lines ½ inch beyond the body.

- shoulder
- chest
- waist
- end of torso
- thigh
- knees
- bottom of feet

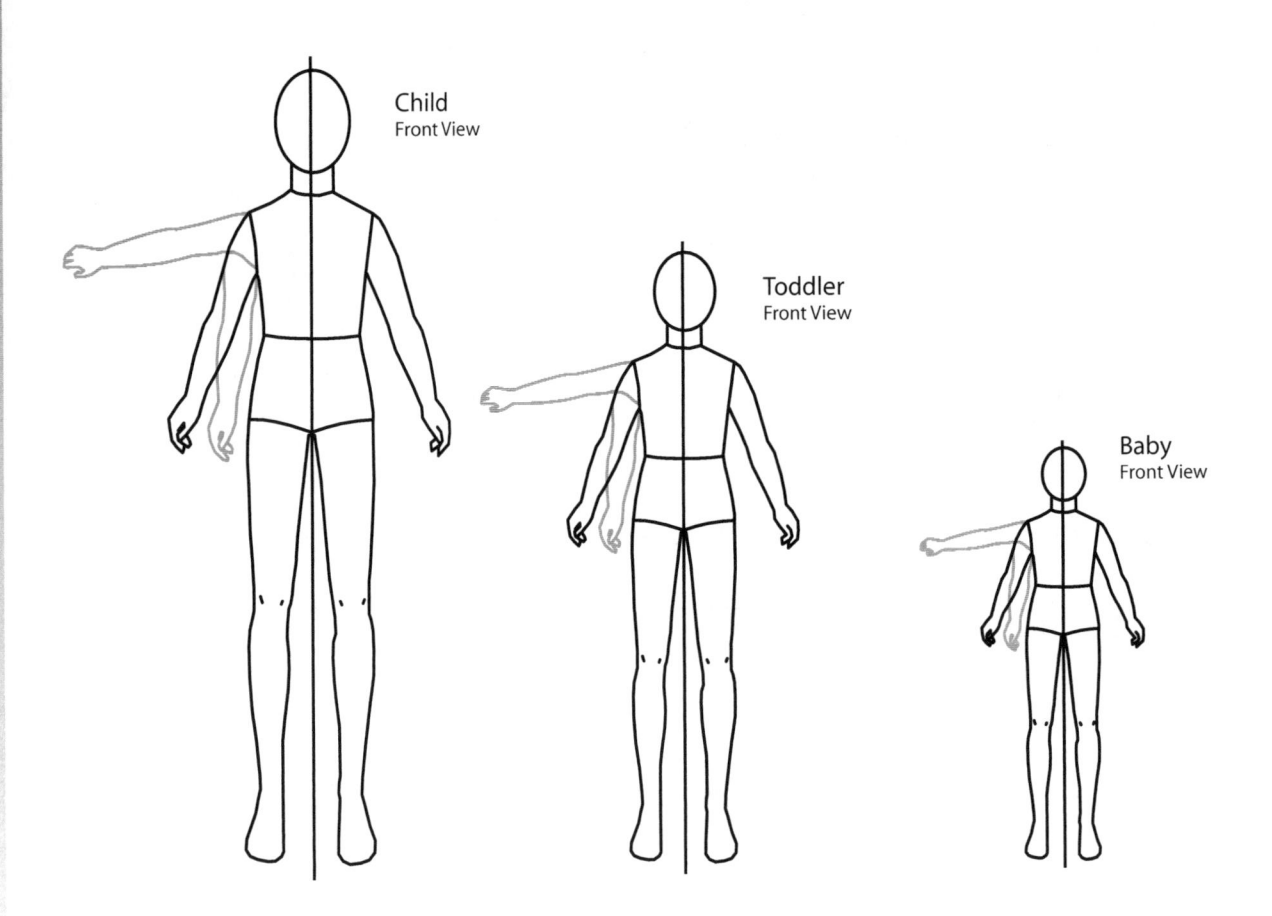

Child
Front View

Toddler
Front View

Baby
Front View

Expeditions in Your Classroom: Geometry © Walch Education

Fashionistas

Off You Go

Activity 1: Fashion Geometry

> **Goal:** To explore the use of geometry and geometric shapes in clothing styles
>
> **Materials:** computer with Internet access, magazines, digital camera (optional)

Is there geometry in what you wear? Find out!

Directions

1. Walk around a mall, look at magazines, or browse your Helpful Web Resources and other sites to find examples of geometric shapes in the cut or patterns of clothing. Go back in time to explore styles from the past.

2. Find at least five examples of geometric shapes in the cut or patterns of clothing. Find at least one example from the past. You are not limited to high fashion. Any type or article of clothing counts.

3. Print out, photocopy, photograph, or draw a very rough sketch of each example.

4. Bring the pictures to class.

Fashionistas

Off You Go

Activity 2: Clothes Inspection

Goal:	To explore clothing construction and dimensions using articles from your own collection
Materials:	tape measure or ruler; an article of clothing
Tools:	Clothes Inspection Worksheet

Directions

1. Choose an article of clothing from your own wardrobe. Bring it to class. Your article should be a shirt, a jacket, a pair of pants, a skirt, or something else you can wear on your torso or lower body (no shoes, belts, hats, underwear, socks, and so forth).

2. In class, examine the construction of the article. Think about the following questions:

    • How many pieces of cloth were sewn to create the article?

    • What geometry do you see in the construction?

    • What would the pattern used to make this piece look like?

3. Use the Clothes Inspection Worksheet to record your findings.

4. Take clothing measurements and record them on your Clothes Inspection Worksheet.

5. Next, make a rough sketch of each piece on your Clothes Inspection Worksheet. Find and mark as many dimensions as you can. This is a reverse engineering challenge—you are working from 3-D to 2-D. For example, a sleeve might look like this and include measurements for each line shown:

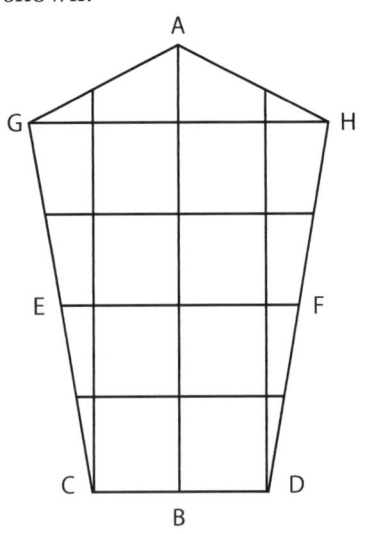

Expeditions in Your Classroom: Geometry

Fashionistas

Expedition Tool

Clothes Inspection Worksheet

Clothing type (shirt, pants, and so forth): _____

Clothing Measurements

Chest: _____

Back to waist: _____

Front shoulder to waist: _____

Back width: _____

Sleeve length: _____

Upper arm: _____

Waist: _____

Hips: _____

Thigh: _____

Skirt length (or waist to knee for pants): _____

Pants length: _____

Head: _____

Pattern Pieces and Dimensions

Sketch the shape of each piece of cloth used in the construction of your clothing article. Find and label as many measurements as you can. Use another sheet of paper if needed.

Fashionistas

Off You Go

Activity 3: You Be the Designer

Goal:	To design an article of clothing and create a graded pattern for it
Materials:	paper, graph paper, pencil, colored pencils or crayons, butcher paper
Tools:	My Design worksheet

It's time to use your emerging fashion and clothing-construction skills on a design of your own—a shirt, a dress, a pair of pants—any type or style you want! You will create a pattern for your design, make it out of paper, and model it.

Directions

1. Brainstorm and create rough sketches of your ideas. Be sure to sketch front and back views.

2. Sketch your final design on your My Design worksheet.

3. Record sewing measurements you need. Base them on your own physical measurements or on those of a friend. If you prefer, you may ask your teacher for a set of measurements you can use.

4. On a sheet of graph paper, create rough pencil sketches of pattern pieces needed for your design. Pay attention to the geometric shapes you use.

5. Calculate and mark any additional measurements you will need for pattern pieces. To start, mark the body measurements you know. Take more measurements if you need them. Use your geometry skills and experience with clothing proportions to calculate or estimate others. You can also drape sample pattern pieces (paper) over a model (you) to help.

   Tips
   - For areas that should fit loosely or allow movement, such as under the arm, add $1/2$ to 1 inch to the measurement. This is called *ease*.
   - Include a seam allowance. The *seam allowance* is the distance between the seam line that joins pieces of fabric together (where you sew) and the edge of the fabric (where you cut).
   - A typical allowance would be $5/8$ inch (1.5 centimeters).

(continued)

Fashionistas

Off You Go

6. Use your sketches and measurements to draw the actual pattern pieces.

    - Draw your pieces on butcher paper (actual scale).
    - Number and label each piece (1—left sleeve, 2—right sleeve, and so forth).
    - Show your stitching line (seam allowance).
    - Don't cut pieces yet!

7. A graded pattern includes lines for larger and smaller versions of the design. Use a different colored pencil to add pattern lines for a size 8% larger OR 8% smaller than your original design size. You may choose larger or smaller.

8. Cut out your pattern pieces. Drape or hold them up to check for fit. Remake a piece if measurements are considerably off.

Fashionistas

Expedition Tool

My Design
Design sketch:

Measurements

Back to waist: _____

Front shoulder to waist: _____

Back width: _____

Chest: _____

Sleeve length: _____

Upper arm: _____

Wrist: _____

Waist: _____

Hips: _____

Thigh: _____

Skirt length (or waist to knee for pants): _____

Pants length: _____

Other: _____

Attach a sheet of graph paper that shows each piece of your pattern, with measurements.

Fashionistas

Off You Go

Activity 4: Paper Clothes Fashion Show

> **Goal:** To construct your article of clothing out of paper and walk it down the runway
>
> **Materials:** butcher paper, tape, glue, art supplies (markers, feathers, glitter, and other decorations)

Directions

1. Create your article of clothing! Calculate the total amount of "fabric" (butcher paper) you need for your pieces. Get that amount from your teacher.

2. Trace your pattern onto your "fabric" and cut out pieces.

3. Tape everything together, as if you were sewing the garment. Use tape sparingly until you are sure the garment is assembled and fitted properly.

4. Use art supplies to style your design. Be creative!

5. Try your creation on and model it for classmates. For fun, create a runway atmosphere, have a panel of "experts" critique designs, let commentators or "fashion police" provide remarks, or give awards (Best Dressed, Fashion Faux Pas, and so forth).

Use your design experiences to help you answer the questions below.

6. How did your knowledge of geometry help you construct your design?

   _____
   _____
   _____
   _____
   _____

7. Which aspects of pattern-drafting were most difficult? Explain.

   _____
   _____
   _____
   _____
   _____

(continued)

Fashionistas

Off You Go

8. Which areas of clothing were most difficult to envision or fit? Explain.

9. Was it difficult to size up and down? Explain.

10. How did the final sizing and proportion of your design work out?

11. Where would more precise measurements or techniques help? Explain.

Fashionistas

Check Yourself!

Skill Check

1. You are making ten gowns for an upcoming theater production. You have a pattern for sizes 8 and 10. What measurements would you use for a size 12? Complete the chart below.

	Size 8	Size 10	Size 12
Chest	31 inches	33 inches	
Waist	22 inches	25 inches	
Hips	29 inches	33 inches	

2. A designer wants to make a pair of pants and a dress using fabric she bought in Milan. She has 2 yards of fabric (1 yard = 3 feet × 3 feet). Her patterns are shown below.

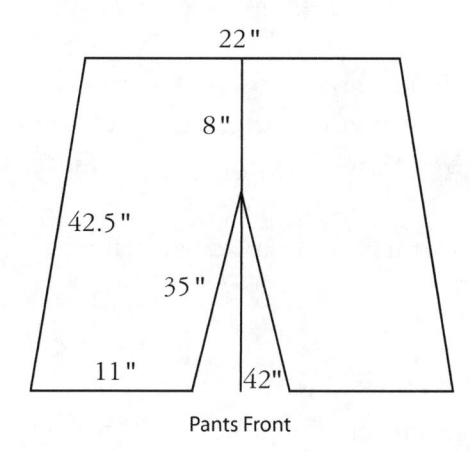

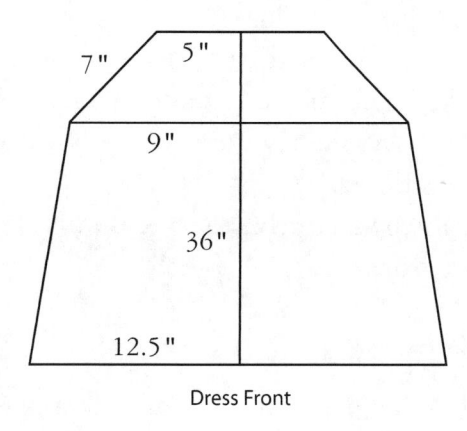

a. Which article of clothing can she make? How much fabric does each require? Explain.

b. For each, how much fabric would she have left or would she need? Explain.

Expeditions in Your Classroom: Geometry

Fashionistas

Check Yourself!

Self-Assessment and Reflection
Project Management

Before You Go

- ❑ I understand the concepts of ratio and proportion and how they apply to clothing design.
- ❑ I'm honestly not sure I understand the math involved in the project and have asked my teacher for additional help.

Off You Go

- ❑ I reviewed the activities and materials for this project and understood products I needed to create.
- ❑ I found at least five examples of geometry in fashion or clothing construction. I was prepared to share pictures or sketches of them in class.
- ❑ I brought an article of clothing from home. I took all requested measurements. I put 100% effort into analyzing its construction, sketching possible pattern pieces, and calculating dimensions.
- ❑ I sketched front and back views of my new design. My ideas are represented clearly.
- ❑ I worked hard to get accurate measurements and figure out pattern piece dimensions.
- ❑ My pieces accurately indicate my original size (with stitching line/allowance) and one additional size.
- ❑ I put my own creative touches on my final garment and participated fully in the runway show.

Do You Know?

- ❑ I can define the Lingo to Learn vocabulary terms for this project and give an example of each.
- ❑ I completed the Skill Check problems and carefully reviewed problems I answered incorrectly.

Fashionistas

Check Yourself!

Reflection

1. What were the most challenging aspects of this project for you and why?

   _____
   _____
   _____
   _____
   _____
   _____
   _____
   _____

2. What skills did this project help you develop?

   _____
   _____
   _____
   _____
   _____
   _____
   _____
   _____

3. If you did this project again, what might you do differently and why?

   _____
   _____
   _____
   _____
   _____
   _____
   _____
   _____

At the Scene of the Crime

Overview
Students invent a mock crime and stage the crime scene. Teams then swap scenes. Students investigate the scene, prepare a crime scene map, and write a reconstruction report.

Time
Total time: 8 to 10 hours

- Before You Go—Crime Scene Reconstruction Crash Course: one 55-minute class period or 45 to 60 minutes of homework
- Before You Go—X Marks the Spot (Crime Scene Coordinates): one 55-minute class period or 45 to 60 minutes of homework
- Activity 1—Design a Crime: two to three 55-minute class periods
- Crime scene setup: 10 minutes in class
- Activity 2—Investigate the Scene: 40 minutes in class
- Activity 3—Crime Scene Report: two to three 55-minute class periods or 2 to 3 hours of homework

Skill Focus
- properties of polygons
- scale drawing
- coordinate systems
- angles
- measurement

Prior Knowledge
- measuring area and other dimensions of polygons
- scale drawing
- graphing
- Some exposure to floor plans is also helpful.

Team Formation
Students work in teams of three to five students.

Lingo to Learn—Terms to Know
- **area:** the number of square units needed to cover a surface
- **perimeter:** the sum of the lengths of the sides of a polygon
- **rectangular coordinates:** Cartesian coordinates where each point on a plane is determined using *x*- and *y*-axis values
- **scale:** the ratio between the size of something and a representation of it

At the Scene of the Crime

- **triangulation:** a method of calculating the location of an object using known measurements of two other objects; creating a triangle from three objects and using side and angle measurements to calculate an unknown measurement

Suggested Steps
Preparation

- Gather any materials and props you may be providing for students.
- Provide administrators and colleagues with an overview of the project. Confirm any areas of the school, types of crime, and so forth that are off-limits or inappropriate.
- Write and send a note home to parents explaining the project.
- Arrange for students to have access to computers and the Internet.

Day 1

1. Form student teams, provide an overview of the project, and show students the project materials.

2. Assign students to work independently or as a team to complete Before You Go: Crime Scene Reconstruction Crash Course.

3. Provide students with computer time and Internet access to research their responses using the project web resources. Have them complete the activity for homework if needed.

Homework

Have students complete Before You Go: Crime Scene Reconstruction Crash Course as needed.

Day 2

1. Assign Before You Go: X Marks the Spot (Crime Scene Coordinates). Have students work in small groups to find measurements.

2. Allocate specific objects in the room, or let each group select two or three objects.

3. As an extension of the project and the math skills covered, show students how to determine paths of trajectory or calculate the angle of impact from splattered or shattered evidence. Consider using the web resources on the following page.

At the Scene of the Crime

- Criminology
  www.schools.utah.gov/cte/documents/wbl/publications/7_8CCC/7Math_Scientific_Criminology.pdf
- Gizmos & Gadgets—Bullet Trajectory Rods
  www.csigizmos.com/products/sceneaccessories/bullettrajectory.html

4. Before the end of class, provide a preview of Activity 1: Design a Crime.

5. Discuss limits you have. Emphasize that students should strive for funny, mysterious, or baffling crimes—not scary, gory, or disgusting ones.

Days 3 and 4

1. Introduce Activity 1: Design a Crime and crime scene criteria.

2. Provide time for teams to meet to plan their crimes. Remind students that you need one legible Design a Crime Worksheet per team. Suggest that each team appoints a recorder.

3. Collect worksheets as teams finish.

4. Inform students of the day the crime scene investigations will take place. Instruct them to begin assembling materials they will need. Note that your final approval is still pending.

5. Review proposed crimes. If time allows, begin your review in class.

6. Make any notes, requests, or comments to support students' understanding. Note any calculations or math that might be unfamiliar to students or pose stumbling blocks.

7. Mark worksheets as "approved" or "needs more work."

Day 5

1. Return Design a Crime Worksheets to students. Highlight any important points from your review.

2. Assign which teams will swap crime scenes. Explain that each team will investigate the other team's scene.

3. Let each team know the location they will investigate or have teams meet quickly to relay this information. No other details of the crime should be discussed.

Expeditions in Your Classroom: Geometry   © Walch Education

At the Scene of the Crime

4. Review the next day's activities. State the following:
   - Two activities will take place during one class—the crime scene setup AND the scene investigation.
   - Students must be ready at the start of class, with the materials they need for both activities ready to go.

Note: Give students at least one day of advance notice to assemble materials and props. Use the in-between time to review problem-solving strategies, any potential stumbling blocks in math that you discovered in your review of the crime scenes, and so forth. Alternatively, give students time to prepare their witness or fine-tune their setup plan.

Day 6 (Crime Day)

1. Define the tasks and time limits for students: 10 minutes to set up and 30 minutes to investigate the other team's scene.
2. Tell each team to appoint a timekeeper. Dispatch teams to set up their scenes. Teams may opt to send only one or two members.
3. After 10 minutes, announce the switch. Dispatch teams to investigate scenes.
4. Call an end to the scene investigations after 30 minutes.
5. Ask one or two members of each team to collect evidence and restore crime scene locations to their original state. Or, members of the investigating teams can do this.

Day 7

1. Review the requirements of Activity 3: Crime Scene Report and the Crime Scene Report Template.
2. Assign the due date of the report. Specify if class time will be used or if students must complete the report as homework.
3. Let teams meet to decide how they will complete assignments. Make suggestions as needed.

Report Due Date

1. Direct teams to review their partner team's report.
2. Tell reviewing teams to write their comments on the final page.
3. Discuss, debrief, and collect reports.

Expeditions in Your Classroom: Geometry

At the Scene of the Crime

Final Day

1. Have students complete the Skill Check problems.
2. Check and review answers.
3. Have students complete the Self-Assessment and Reflection worksheet and submit it (optional).

Project Management Tips and Notes

- The crime scene setup/investigation activities are best suited to an extended or block period. Alternatively, direct students to select specific crime scene locations of your choosing that can remain undisturbed from one day to the next so setup can be done in advance.
- Some students may struggle to develop crime scene sketches. Show students a floor plan example. Remind them that they can use basic shapes and symbols to represent items. Explain some of the symbols commonly used (for doors, windows, and so forth).
- To constrain the project, construct one scene of your own design for all teams to investigate. If you do, ask students to brainstorm and submit ideas for you to use.

Suggested Assessment

Use the Geometry Project Assessment Rubric or the following point system:

Team and class participation	10 points
Crime scene design	10 points
Crime scene investigation	15 points
Crime scene report	60 points
Project self-assessment	5 points

Extension Activities

- Have students present their findings as crime scene investigators testifying in court.
- Create crime scene floor plans using a computer-aided design (CAD) tool.
- Design an interdisciplinary project with science classes (physics for projectiles, biology for forensics, and so forth).

At the Scene of the Crime

Common Core State Standards Connection

High School

Geometry: Modeling with Geometry

G-MG.1. Use geometric shapes, their measures, and their properties to describe objects (e.g., modeling a tree trunk or a human torso as a cylinder).★

G-MG.3. Apply geometric methods to solve design problems (e.g., designing an object or structure to satisfy physical constraints or minimize cost; working with typographic grid systems based on ratios).★

Number and Quantity: Quantities★

N-Q.1. Use units as a way to understand problems and to guide the solution of multi-step problems; choose and interpret units consistently in formulas; choose and interpret the scale and the origin in graphs and data displays.

Grade 8

Geometry

8.G.6. Explain a proof of the Pythagorean Theorem and its converse.

8.G.8. Apply the Pythagorean Theorem to find the distance between two points in a coordinate system.

Grade 7

Geometry

7.G.1. Solve problems involving scale drawings of geometric figures, including computing actual lengths and areas from a scale drawing and reproducing a scale drawing at a different scale.

Other Helpful Resources

- truTV—Forensics in the Classroom
  http://apps.trutv.com/forensics_curriculum/

- Microsoft Office Visio—Create a Floor Plan
  http://office.microsoft.com/en-us/visio-help/create-a-floor-plan-HP001208559.aspx

Expeditions in Your Classroom: Geometry

At the Scene of the Crime

Answer Key

Before You Go: Crime Scene Reconstruction Crash Course

1. Answers will vary but may include interviewing first responders; surveying the scene (define the extent and boundaries of the scene and secure the area); getting search warrants; doing an initial walk-through and taking notes (preliminary survey); calling in specialists if needed; searching the scene (detailed examination, evidence collection, and documentation); doing a final survey; releasing the scene.

2. Forensics relates to work done back at the lab and involves the analyzing of evidence collected. Crime scene reconstruction involves documenting the scene and creating visual records of it.

3. Patterns include inward spiral, outward spiral, parallel search, grid search, and zone search (all three-dimensional).

4. overview shots, mid-range view, close-ups of individual pieces of evidence

5. An investigator might take a second set of photographs with a ruler or other common object (pen, car key) alongside the item to show scale.

6. A sketch can span the entire scene (for example, several rooms) and record aspects of the scene with exact measurements. A sketch artist would be sure to record the size of the room, the dimensions of possible exits and entrances such as doors and windows, and the location of evidence and how each piece relates to the rest of the scene and other evidence.

7. the exact spot where the offense took place; areas from which the site can be entered, exited, or escaped; locations of key pieces of evidence

8. Answers will vary. Possible answers:
   You rarely see investigators waiting around for a search warrant; analysis of evidence doesn't usually happen as quickly or as accurately; it's very difficult to pin down time of death to a two-hour range; crime scene investigators do not usually interview witnesses or suspects; a crime scene investigator rarely handles an entire investigation since he or she usually doesn't have the expertise or time to do it all.

9. Answers will vary. Possible answers:
   to sketch the scene and make scale drawings; to determine footprint sizes; to figure out the angle of impact of a bullet or a blow; to figure out trajectory; to determine the distance between objects and evidence

Check Yourself! Skill Check

1. a. 18 inches
   b. 9 inches
   c. 23°
   d. 67°

At the Scene of the Crime

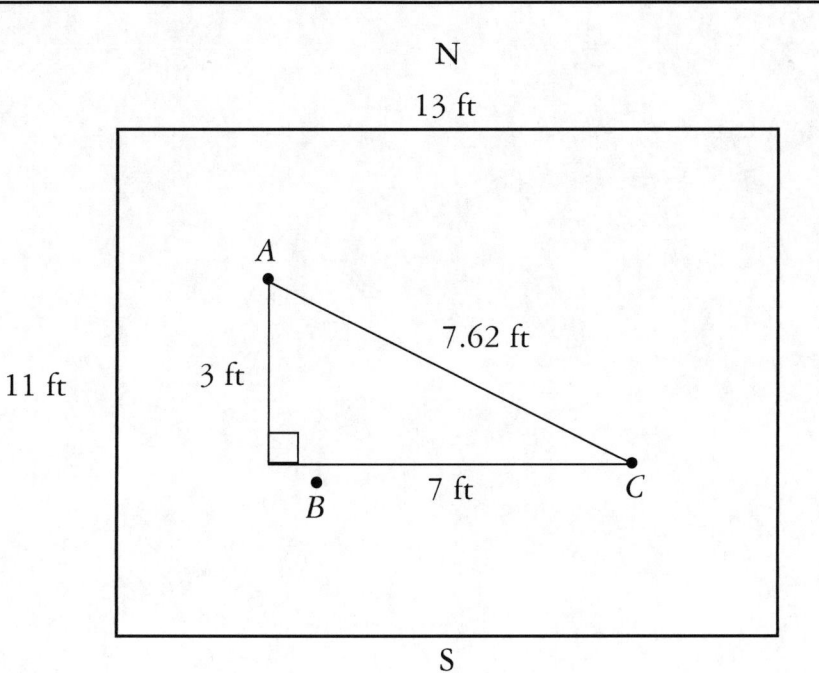

N
13 ft

A
7.62 ft
3 ft
11 ft
B 7 ft C

S

Triangulation

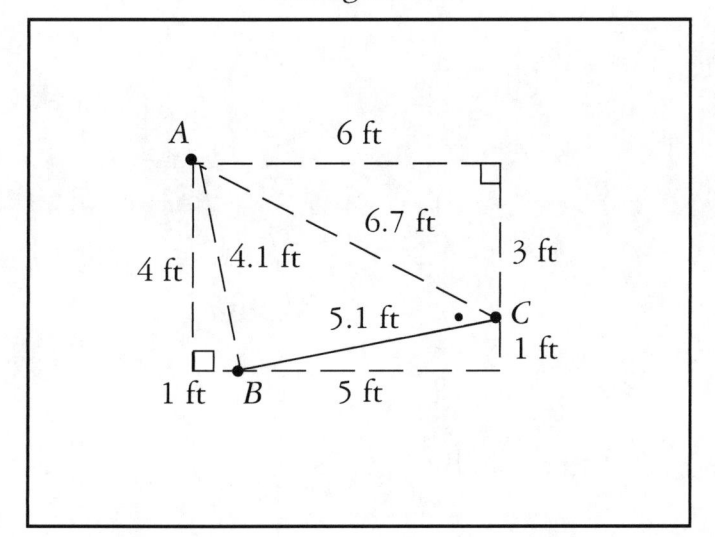

A 6 ft
6.7 ft
4 ft 4.1 ft 3 ft
5.1 ft C
1 ft
1 ft B 5 ft

2. a. $a^2 + b^2 = c^2$

$3^2 + 7^2 = c^2$

$9 + 49 = c^2$

$c^2 = 58$

$c = 7.62$ feet

At the Scene of the Crime

b. Using triangulation:

Distance from *A* to *B*:

$4^2 + 1^2 = c^2$

$16 + 1 = c^2$

$c = 4.1$

Distance from *B* to *C*:

$a^2 + b^2 = c^2$

$1^2 + 5^2 = c^2$

$1 + 25 = 26$

$c = 5.1$ feet

Distance from *A* to *C*:

$3^2 + 6^2 = c^2$

$9 + 36 = c^2$

$c = 6.7$ feet

The evidence is 4.1 feet from *B* and 6.7 feet from *A*.

At the Scene of the Crime

Expedition Overview

Challenge
Your school is experiencing a wave of peculiar crimes. Not to worry! Your class has the brains, talent, determination, and math skills needed to get to the bottom of things.

For this project, you will play the roles of a criminal mastermind and a crime scene investigator. As a crime scene investigator, you will be part of a team that will analyze a scene created by another team and report your findings. Grab your measuring tape and protractor, and get ready to catch a crook!

Objectives
- To use coordinate geometry to describe locations
- To reconstruct the details of a floor plan—and crime scene evidence—in a scale drawing
- To apply geometric concepts and calculations to real-world situations

Project Activities
Before You Go
- Crime Scene Reconstruction Crash Course
- X Marks the Spot (Crime Scene Coordinates)

Off You Go
- Activity 1: Design a Crime
- Activity 2: Investigate the Scene
- Activity 3: Crime Scene Report

Expedition Tools
- Design a Crime Worksheet
- Crime Scene Checklist
- Evidence Inventory
- Witness Statement
- Crime Scene Report Template

Other Materials Needed
- notebook
- paper or graph paper
- pens and pencils
- ruler
- tape measure or roll of string
- protractor

At the Scene of the Crime

Expedition Overview

- compass
- calculator
- props for crime scene "evidence" (shattered objects, fake blood, articles of clothing, and so forth)
- digital camera (optional)

Lingo to Learn—Terms to Know

- area
- perimeter
- rectangular coordinates
- scale
- triangulation

Helpful Web Resources

- Crime Scene Investigation
  www.crime-scene-investigator.net

- Criminology
  www.schools.utah.gov/cte/documents/wbl/publications/7_8CCC/7Math_Scientific_Criminology.pdf

- Dummies.com—Forensics: Assessing the Scene of the Crime
  www.dummies.com/how-to/content/forensics-assessing-the-scene-of-the-crime.html

- FBI Laboratory Services—Forensic Science Support
  www.fbi.gov/about-us/lab/forensic-science-support

- Federal Bureau of Investigation—Famous Cases & Criminals
  www.fbi.gov/libref/historic/famcases/famcases.htm

- Federal Bureau of Investigation—Handbook of Forensic Services
  www.fbi.gov/about-us/lab/handbook-of-forensic-services-pdf
  (See the "Crime Scene Search" section beginning on page 171 of this PDF.)

- Florida State University College of Criminology and Criminal Justice—Crime Scene Measurement
  http://criminology.fsu.edu/faculty/nute/CSmeasurement.html

- Gizmos & Gadgets—Bullet Trajectory Rods
  www.csigizmos.com/products/sceneaccessories/bullettrajectory.html

Expeditions in Your Classroom: Geometry

At the Scene of the Crime

Expedition Overview

- How Stuff Works—How Crime Scene Investigation Works
  http://science.howstuffworks.com/csi.htm

- Microsoft Office Visio—Create a Floor Plan
  http://office.microsoft.com/en-us/visio-help/create-a-floor-plan-HP001208559.aspx

- truTV
  www.trutv.com/index.html

- truTV—Forensics in the Classroom
  http://apps.trutv.com/forensics_curriculum/

At the Scene of the Crime

Before You Go

Crime Scene Reconstruction Crash Course

> **Goal:** To learn about crime scene investigation methods and scene reconstruction
>
> **Materials:** computer with Internet access

Directions
Use your Helpful Web Resources to research crime scene investigation and answer the questions below.

1. What steps does a crime scene investigator take to process a scene?

   _____
   _____
   _____

2. What is the difference between forensics and crime scene reconstruction?

   _____
   _____
   _____

3. What patterns might a crime scene investigator use to search a crime scene?

   _____
   _____
   _____

4. Typically, what views would a crime scene investigator sketch or photograph at a crime scene?

   _____
   _____
   _____

(continued)

At the Scene of the Crime

Before You Go

5. In photographs, how might a crime scene investigator indicate the scale of evidence found?

   _____
   _____
   _____

6. Why is it important to create sketches of a scene (instead of only using photos and video)? What important details might a sketch artist note?

   _____
   _____
   _____

7. At a minimum, what areas does a crime scene include?

   _____
   _____
   _____

8. Television shows and movies do not always portray the science and art of crime scene investigation accurately. Describe at least two examples of common inaccuracies.

   _____
   _____
   _____

9. How do you think math skills relate to crime scene investigation? How might an investigator use geometry and math skills? Give examples.

   _____
   _____
   _____

At the Scene of the Crime

Before You Go

X Marks the Spot (Crime Scene Coordinates)

> **Goal:** To explore measurement techniques used for crime scene investigation
>
> **Materials:** tape measure

There are two techniques crime scene investigators use to take scene measurements and identify the location of evidence and other important information.

Triangulation

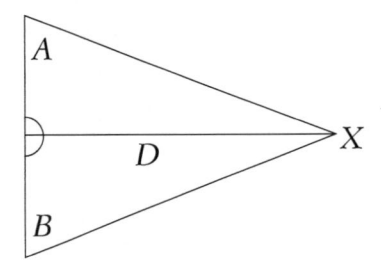

To use triangulation to determine location, select two fixed points at the scene and measure the distance between them. Measure the distance or angle of an object from both points.

Rectangular Coordinates

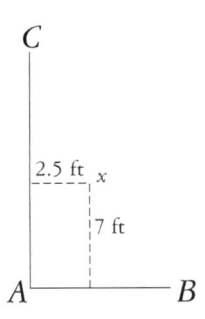

You can use rectangular coordinates to determine a location by selecting two fixed points and creating a reference line with string or tape (the baseline approach). Measure along the line and perpendicular to the line. Alternatively, measure from two perpendicular baselines (the grid approach).

(continued)

Expeditions in Your Classroom: Geometry

At the Scene of the Crime

Before You Go

Directions

Choose three or four objects in your classroom. Practice giving locations of the objects using each technique.

Item	Location using triangulation	Location using rectangular coordinates

Expeditions in Your Classroom: Geometry

At the Scene of the Crime

Off You Go

Activity 1: Design a Crime

> **Goal:** To brainstorm, detail, and stage a "crime" (with your team) to be investigated by another team
>
> **Materials:** pencil, paper, ruler, measuring tape or string, "evidence" props
>
> **Tools:** Design a Crime Worksheet

Directions

1. With your team, review the crime scene criteria below and brainstorm crime ideas.

> **Crime Scene Criteria**
>
> ❑ Your crime should be clever and creative. However, this is an educational adventure, not meant to be gory or to spread panic. Use common sense. Your crime should be fictitious. Obviously, you will not do anything illegal.
>
> ❑ Your crime must occur in school locations approved by your teacher.
>
> ❑ Provide at least one witness—a member of your team who is prepared to answer questions posed by the investigating team that might help them solve the crime.
>
> ❑ Leave a minimum of three items of evidence at the scene that relate directly to the perpetrator or perpetration of the crime.
>
> ❑ Evidence must be located on multiple planes (floor, wall, ceiling, and so forth) in order to challenge investigators' skills in coordinate math, measurement, and three-dimensional drawing.
>
> ❑ You must be able to set up your crime scene in 10 minutes or less.

2. Choose your best idea and formulate a detailed, step-by-step plan and story line for your crime. What crime occurred and how?

3. Record the details of your plan on the Design a Crime Worksheet.

4. Prepare your witness or witnesses.

5. Following your teacher's instructions, stage your crime at the location you selected.

At the Scene of the Crime

Expedition Tool

Design a Crime Worksheet

Your team name: _____

Your crime scene location: _____

1. What will the crime scene location look like? Attach a detailed sketch that shows important features and objects (walls, doors, windows, furniture, and so forth) and their dimensions. Include accurate labels and measurement information. Your drawing does not need to be done to scale.

2. What is the crime and how will it happen? Write a step-by-step description of how the crime will occur.

   _____
   _____
   _____
   _____
   _____

3. Make a list of the evidence perpetrators will leave behind. What, if anything, about the location will change as a result of the crime? You can provide location information relative to a baseline, another object in the room, or another piece of evidence.

   Mark the location of evidence on your crime scene sketch.

Item	Description	Location	Measurements

(continued)

At the Scene of the Crime

Expedition Tool

4. Is there anything else important to know about the scene (for example, lighting conditions or angle of lighting, security, time of day, used or vacant, and so forth)?

   _____
   _____
   _____

5. What will your witness(es) see and when? Invent a story for each witness and mark the location or vantage point of each on your crime scene sketch.

   - Who is the witness?

   - When was he or she at the scene of the crime?

   - Why was he or she at the scene of the crime?

   - What are his or her physical characteristics?

   - What was his or her vantage point or field of view?

   - What did/could the witness see? What didn't the witness see and why?

   - What else does the witness know or not know about the crime scene?

6. List any materials or props will you need to stage your crime.

   _____   _____
   _____   _____
   _____   _____

Expeditions in Your Classroom: Geometry

At the Scene of the Crime

Off You Go

Activity 2: Investigate the Scene

Goal:	To investigate and document another team's crime scene
Materials:	notebook, sketch paper, pencils, digital camera, ruler, measuring tape or string, compass, protractor
Tools:	Crime Scene Checklist, Evidence Inventory, Witness Statement

Directions

1. Organize your team and assemble the materials you need to document the scene. All team members should take notes. Assign specific roles and duties to each person; for example, there should be a lead investigator, a sketch artist, a photographer, an evidence recorder, a witness interviewer, and so forth. You may need to play more than one role.

2. Go to the scene of the crime. As a team, determine your approach.

   - What are your initial impressions? Do you have any hunches or theories?
   - What are the boundaries of the crime scene?
   - How will you do your walk-through?
   - How will you get the measurements you need?

3. Use the Crime Scene Checklist and the Witness Statement tools to help you conduct your search, collect data, and document the scene. Remember, according to the FBI, physical evidence cannot be overdocumented!

4. Interview witnesses at any stage of your crime scene search. Witnesses may be interviewed once, and for no more than five minutes. You may interview only one witness at a time.

5. When your investigation is complete, compare notes as a team and do a final survey of the scene.

At the Scene of the Crime

Expedition Tool

Crime Scene Checklist

Use this checklist to help you collect the information you need during your crime scene search.

- ❏ Investigator names
- ❏ Type of crime
- ❏ Crime scene location
- ❏ Date and time arrived at scene
- ❏ Scene conditions
- ❏ Search approach and methods used
- ❏ Description of the crime scene and search findings
- ❏ Measurements (scene area and boundaries; location and size of evidence; other relevant measurements such as victim's height, shoe size, size of windows and doors, and so forth)
- ❏ List of sketches (overview and side-view shots required; others optional as needed)
- ❏ List of photos (optional)
- ❏ List of witnesses interviewed/witness statements
- ❏ Evidence inventory

At the Scene of the Crime

Expedition Tool

Evidence Inventory

List key items of evidence discovered at the scene.

Item	Description	Location	Measurements

Expeditions in Your Classroom: Geometry © Walch Education

At the Scene of the Crime

Expedition Tool

Witness Statement

Witness name: _____

Interviewed by: _____

Relationship to victim or other witnesses: _____

Profile (age, occupation, relevant physical characteristics, and so forth):

_____
_____
_____
_____
_____
_____

Statement:

_____
_____
_____
_____
_____
_____
_____
_____
_____
_____
_____
_____
_____
_____

At the Scene of the Crime

Off You Go

Activity 3: Crime Scene Report

Goal:	To create a crime scene report that includes sketches and analysis
Materials:	pencils, ruler, protractor, computer
Tools:	Crime Scene Report Template

Directions

1. With your team, review the Crime Scene Report Template. Decide how you will organize as a team to prepare and integrate components of the report.

2. Review and analyze your team's findings. Develop a hypothesis that reconstructs the crime and the steps taken to commit it. What evidence supports your theories?

3. Prepare a draft of your Crime Scene Report. Review the draft as a team and make any necessary revisions or corrections.

4. Type your final report. Sketches should be attached and may be hand-drawn, computer-generated, or a combination of both.

5. Give your report to the team that designed the crime you investigated.

6. Comment on the report submitted to your team. How accurate are the report and conclusions presented by the investigating team?

7. Submit your final report to your teacher.

At the Scene of the Crime

Expedition Tool

Crime Scene Report Template

> Case: _____
>
> Date: _____
>
> Nature of offense: _____
>
> Investigating officers: _____

Date and time of scene search: _____

Location and scene description: _____

_____

Summary of Search Findings

Describe your search. Write four to five paragraphs detailing what you observed and discovered at the scene. Use your own paper if you need more space.

_____
_____
_____
_____
_____
_____
_____
_____
_____
_____
_____

(continued)

At the Scene of the Crime

Expedition Tool

_____
_____
_____
_____

Evidence Log

Item	Description	Location	Measurements

List all evidence found or collected.

Scene Sketches

On separate sheets of paper, prepare a minimum of two scale drawings of the crime scene—one overview or floor-plan view and one side view. You may include additional sketches and photos if you want.

(continued)

Expeditions in Your Classroom: Geometry

At the Scene of the Crime

Expedition Tool

Crime Scene Drawings

Your sketches should include the following:

- ❏ keys or legends to indicate direction and scale (using either metric or English measurements)
- ❏ dimensions of the room or rooms (length, width, height)
- ❏ location and dimensions of doors, windows, or other suspected entrances and exits
- ❏ location and dimensions of important objects (shelves, desk, lockers etc.)
- ❏ location and position of the victim
- ❏ location of evidence using triangulation or rectangular coordinates (your teacher may specify)
- ❏ location or vantage point of witnesses
- ❏ any other details you feel are important or relevant (room numbers, carpeting on the floor, source of light, etc.)

If studied/requested by your teacher, also include the following:
- ❏ angle of impact, if paint or fluid splatters found
- ❏ trajectory or path, if a projectile is involved

Briefly profile any witnesses and summarize any relevant and credible information they provided.

_____
_____
_____
_____
_____
_____
_____

(continued)

At the Scene of the Crime

Expedition Tool

Conclusion and Recommendations

In three or four paragraphs (on a seperate sheet of paper), give your theory of how the crime occurred, who committed the crime (if you have suspects), and how evidence found at the scene supports your conclusions. Indicate your level of certainty and provide recommendations for further investigation.

Investigator Signatures

_____

_____

_____

_____

Reviewer Comments

Include a blank page for comments from the team that originally staged the crime. They will review your report and conclusions.

At the Scene of the Crime

Check Yourself!

Skill Check

1. Use the figure to find the measures below.

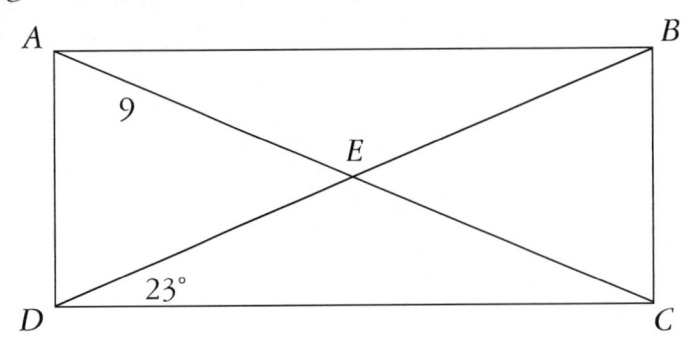

   a. *BD* _____

   b. *EC* _____

   c. ∠*DBA* _____

   d. ∠*ACB* _____

2. At the scene of an art museum heist, you find a piece of evidence, *A*, near two large marble sculptures, *B* and *C*.

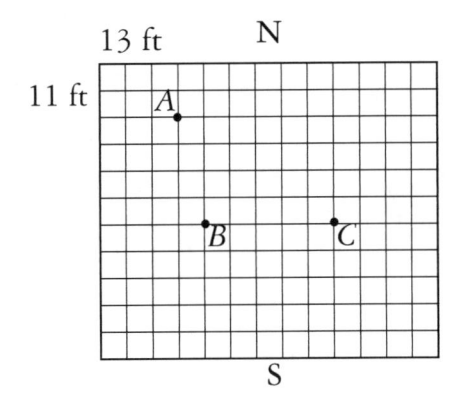

   a. How far is the item of evidence from sculpture *C*?

   b. Describe the location of the evidence using triangulation.

At the Scene of the Crime

Check Yourself!

Self-Assessment and Reflection
Project Management

Before You Go

- ❏ My answers to Before You Go: Crime Scene Reconstruction Crash Course are complete and based on good research.
- ❏ I completed Before You Go: X Marks the Spot (Crime Scene Coordinates) and understand the math skills I need for this project.
- ❏ I'm honestly not sure I understand the math involved and have asked my teacher for additional help.

Off You Go

- ❏ I reviewed our project challenge and project materials carefully and thoroughly. I understand the scope and requirements of the project and final Crime Scene Report.
- ❏ I reviewed the Crime Scene Criteria before brainstorming crime ideas.
- ❏ I contributed substantively to our crime idea and plan.
- ❏ I helped my team divide work evenly and performed all roles and duties assigned to me in a timely manner.
- ❏ I took detailed, legible notes during our crime scene search.
- ❏ I gave my team my full cooperation and participated actively during crime setup and investigation activities.
- ❏ I made significant, high-quality contributions to our final Crime Scene Report.
- ❏ The report is well-organized and detailed. It includes all of the information outlined in the report template. It provides evidence-based supporting points for the final conclusion and recommendations.
- ❏ I reviewed the complete report and am prepared to present any part of it.
- ❏ My team reviewed another team's report and responded with thoughtful, constructive comments.

Do You Know?

- ❏ I can define the Lingo to Learn vocabulary terms for this project and give an example of each.
- ❏ I completed the Skill Check problems and carefully reviewed problems I answered incorrectly.

At the Scene of the Crime

Check Yourself!

Reflection

1. What were the most challenging aspects of this project for you and why?

2. What skills did this project help you develop?

3. If you did this project again, what might you do differently and why?

Protectors of the Realm

Overview
Students plot the locations of British and Welsh castles and create Voronoi diagrams to define the boundaries of their castle "territory" and solve other related problems.

Time
Total time: 5 to 8 hours

- Before You Go—That's Very Voronoi: one 55-minute class period
- Activity 1—Castles Map: one to two 55-minute class periods or homework assignments
- Activity 2—The Extent of Your Domain: one to two 55-minute class periods or homework assignments
- Activity 3—Wise Counsel: two to three 55-minute class periods or homework assignments

Skill Focus
- properties of polygons
- graphing/coordinate systems
- triangulation
- geometric problem solving

Prior Knowledge
- map reading (identifying longitude and latitude coordinates, calculating distance from point A to point B)
- graphing (plotting graph coordinates, calculating slope)
- perpendicular bisectors
- finding perimeter and area of polygons

Team Formation
Students work in teams of two to four students for eight to ten castle locations. This can also be an individual project with students working in groups on Voronoi diagrams.

Lingo to Learn—Terms to Know
- **angle bisector:** a ray/line that divides an angle into two congruent angles
- **circumcenter:** the center of the smallest circle that can contain the shape completely within it
- **edge:** the side of a polygon or a polyhedron

Expeditions in Your Classroom: Geometry © Walch Education

Protectors of the Realm

- **incenter:** the center of a triangle identified using an incircle (the center of the largest circle that can be formed within the triangle)
- **perpendicular bisector:** the line that is perpendicular to a segment at its midpoint
- **polygon:** a closed planet figure formed by line segments that do not intersect
- **vertex:** corner; a point where lines, rays, or polygon sides or polyhedron edges meet
- **Voronoi diagram or tessellation:** decomposition of a plane where the set of points within a cell is closer to a point *A* than to any other point in the plane diagram of cells

Suggested Steps

Preparation

- Make copies of a map of England and Wales (8 $\frac{1}{2}$" × 11" or larger). Print one copy per team and several extras.
- Gather materials as needed, including one ruler per team (or yardstick if you use an enlarged map).
- Consider how to group students into teams.
- Brush up on how Voronoi polygons are used in everyday life to define the area of influence or "zones of proximity" around points in a plane. Voronoi diagrams, also known as Thiessen polygons, are very useful in situations where boundaries are otherwise undefined. Be prepared to explain some of these applications to students:
  - Urban planning and zoning: planning catchment areas for schools, library branches, fire stations, and other public services
  - International development: identifying proximity of villages to resources such as water or hospitals
  - Animal behavior: analyzing an animal's roaming or hunting territory (water and food sources)
  - Business: identifying sales territory or franchise locations
  - Economics: modeling market areas or regional economic influence
  - Medicine: identifying possible sources of an illness or disease affecting a population

Day 1

1. Form student teams, give an overview of the project, and review project materials.

2. Introduce Voronoi diagrams and facilitate Before You Go: That's Very Voronoi.

3. After students complete the activity, ask them to brainstorm possible applications. If needed, give them hints (for instance, ask who might care about "territory" or "districts").

Expeditions in Your Classroom: Geometry

Protectors of the Realm

4. Have teams choose a castle name from a cup or a hat. Explain that castles are actual medieval castles built in England or Wales.

5. Assign homework. Emphasize that teams have two tasks: find out interesting information about the castle, and most importantly, find the location of the castle in terms of longitude and latitude. Encourage students to take notes because they will be asked to share information about their castle in class.

Homework

Have students research their castle location and history.

Day 2

1. Provide additional time to research castles if needed.

2. Distribute one England/Wales map to each team.

3. In class, have each team identify their castle, locate it on a map, name the coordinates, and describe anything interesting they learned. Give teams 3 to 5 minutes each.

4. Make sure that all teams record the information and mark the location of each castle on their team map.

5. Direct teams to compare maps. All maps should be identical.

Day 3

1. Tell teams to take out their England/Wales map.

2. Make sure students mark the outer boundaries or edge of the plane for their Voronoi diagram. Give students the following information—latitude: N51° 45'; latitude: N53° 30'; longitude: W4° 30'; longitude: E1° 30'.

3. Direct teams to start creating their Voronoi diagram on the map. Students should first draw lines to connect castle points.

4. Discuss ideas for how to tackle the problem.

5. After students have drawn at least one polygon, stop them and ask several teams to show their map.

6. If there is time, continue the following day. Otherwise, assign students to complete maps for homework (make copies for each student and divide areas.)

Expeditions in Your Classroom: Geometry © Walch Education

Protectors of the Realm

Day 4

1. Once students finish complete diagrams, discuss the results.

2. Facilitate a comparison of the maps to ensure consistent results. Use the discussion questions provided in Activity 2: The Extent of Your Domain as a jumping-off point. If team maps differ, identify why and make adjustments.

3. Set up Activity 3: Wise Counsel.

4. Review final product requirements listed in the activity instructions.

5. Highlight the geometry problems involved. Discuss how students might approach them.

6. Assign a date for the final products and presentation. Specify if class time will be used or if students must complete the assignments as homework.

Homework or Two Class Days

Have students work on the final product.

Presentation Day

1. Create a festive atmosphere for fun (have trumpet fanfare, use medieval lingo, and so forth).

2. Allow each team five minutes to present.

3. Collect recommendations reports (final products).

Final Day

1. Ask students to complete the Skill Check problems.

2. Check and review answers.

3. Ask students to complete the Self-Assessment and Reflection worksheet and submit it (optional).

Protectors of the Realm

Project Management Tips and Notes
- A diagram with 8 to 10 points can take 1 to 2 hours to complete. The more points involved, the more complicated—and tedious—it is to create a Voronoi diagram. If this is a concern, form larger teams and assign fewer castle locations.
- Rather than having students plot castle points on the map, you can provide a map that already has them. If so, you might also provide a larger map. This makes it easier for teams to see and analyze territories.

Suggested Assessment
Use the Geometry Project Assessment Rubric or the following point system:

Team and class participation	10 points
Castle map (Voronoi diagram)	25 points
Recommendations report	50 points
Recommendations presentation	10 points
Project self-assessment	5 points

Extension Activities
- Discuss other applications of Voronoi diagrams.
- Use GIS software to generate a map with coordinates and Voronoi polygons.
- Ask students to look at the geometry in the architecture of their castle.

Common Core State Standards Connection
High School
Geometry: Modeling with Geometry

G-MG.1. Use geometric shapes, their measures, and their properties to describe objects (e.g., modeling a tree trunk or a human torso as a cylinder).★

G-MG.3. Apply geometric methods to solve design problems (e.g., designing an object or structure to satisfy physical constraints or minimize cost; working with typographic grid systems based on ratios).★

Number and Quantity: Quantities★

N-Q.1. Use units as a way to understand problems and to guide the solution of multi-step problems; choose and interpret units consistently in formulas; choose and interpret the scale and the origin in graphs and data displays.

Protectors of the Realm

Grade 8

Geometry

8.G.9. Know the formulas for the volumes of cones, cylinders, and spheres and use them to solve real-world and mathematical problems.

Grade 7

Geometry

7.G.1. Solve problems involving scale drawings of geometric figures, including computing actual lengths and areas from a scale drawing and reproducing a scale drawing at a different scale.

Answer Key
Before You Go: That's Very Voronoi

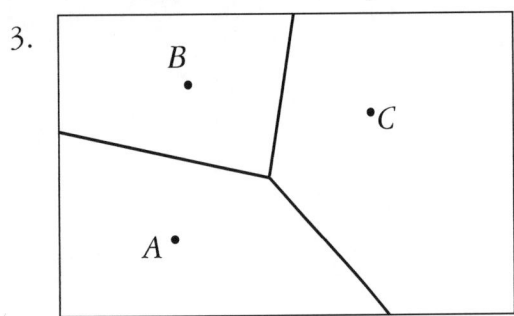

3.

Check Yourself! Skill Check

1. pride *F*
2. hole *E*
3. See diagram below.
4. prides *F* and *D*
5. pride *F*; 23 square miles
6. At the Voronoi vertex; the approximate coordinates on the grid are (9, 9). Prides *B* and *D* would have to travel approximately 1 mile to the water hole. Prides *C* and *F* would have to travel approximately $1\frac{1}{4}$ miles.

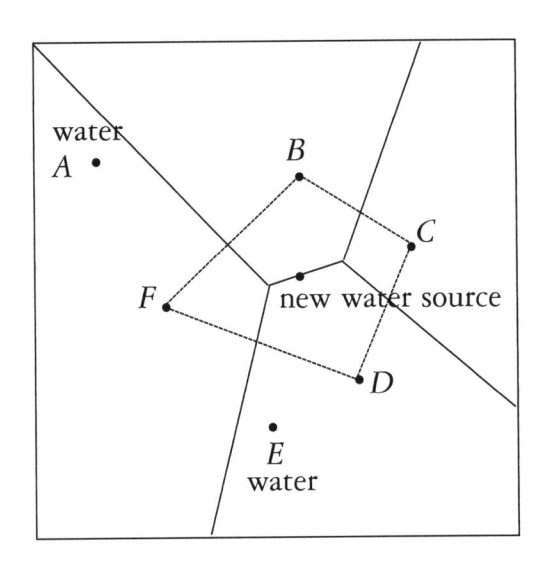

Protectors of the Realm

Expedition Overview

Challenge

You are the trusted adviser of a medieval feudal lord who, due to his loyal service to the king, has been granted a castle stronghold and an expansive amount of land. Per the king's orders, the lord's territories extend as far as a bird can fly in all directions and not meet a bird of the same speed dispatched at the same time from another castle or until it reaches the end of the realm. This is your district to help steward and protect.

Your liege has asked you to evaluate his territory. Among other things, he needs to know how much territory he has, any important geographic features, physical assets it contains, and the locations and boundaries of other castle territories (potential allies or enemies). Most importantly, he wants you to recommend improvements that will help him effectively manage and defend the territory.

Objectives
- To model geographic locations and territories using Cartesian coordinates
- To learn how to draw a Voronoi diagram and use the method to define the "zone of proximity" for a single point
- To understand the properties of polygons and how to calculate measurements

Project Activities
Before You Go
- That's Very Voronoi

Off You Go
- Activity 1: Castles Map
- Activity 2: The Extent of Your Domain
- Activity 3: Wise Counsel

Expedition Tools
Medieval Castles in England and Wales

Other Materials Needed
- enlarged outline map of the United Kingdom
- atlas or access to Google Maps or Google Earth
- graph paper
- paper
- colored pencils or crayons
- ruler
- compass
- protractor

Protectors of the Realm

Expedition Overview

Lingo to Learn—Terms to Know
- angle bisector
- circumcenter
- edge
- incenter
- perpendicular bisector
- polygon
- vertex
- Voronoi diagram or tessellation

Helpful Web Resources
- About.com—Geography: Blank Outline Map of United Kingdom
  http://geography.about.com/library/blank/blxuk.htm

- American Mathematical Society—Voronoi Diagrams and a Day at the Beach
  www.ams.org/featurecolumn/archive/voronoi.html

- BBC—h2g2 Mathematics: Thiessen Polygons
  www.bbc.co.uk/dna/h2g2/A901937

- Google Earth
  http://earth.google.com

- Google Maps
  http://maps.google.com/maps?tab=wl

- National Council of Teachers of Mathematics—Illuminations: Hospital Locator
  http://illuminations.nctm.org/ActivityDetail.aspx?ID=156

- TimeRef—Castles of England and Wales
  www.timeref.com/map3.htm

Protectors of the Realm

Before You Go

That's Very Voronoi

> **Goal:** To learn what a Voronoi diagram is and how it is used as a geometric modeling method
>
> **Materials:** paper, ruler, compass

Directions

1. How would you define your "personal space"? Are the desks next to you occupied by students? If not, does this mean you have more personal space today? As a class, suggest methods you could use to define *your* area or zone.

2. Draw a graph representing the following scenario or, at your teacher's direction, act out the scene:

   - There are six students seated in different locations in the room.
   - They all stand. Each student turns to another standing student to create a pairing.
   - All students walk directly toward their partner at the same pace. Partners meet at some point in the middle. The floor is marked at the meeting point.
   - Everyone returns to their desks.
   - This time, each student turns to face a different person to partner with.
   - Again, the partners walk toward one another. The floor is marked where they meet. Everyone returns to their starting points.
   - The process is repeated until all six students have met one another in this manner.

3. If you were to view one student from above and connect the dots he or she marked on the floor, what shape would you have?

4. You have just created a Voronoi diagram. Propose a hypothesis about how to define the "zone" around each student. Could you create the same diagram if you couldn't see students walking and marking spots on the floor? How?

Expeditions in Your Classroom: Geometry

Protectors of the Realm

Before You Go

5. Test your hypothesis. What is the Voronoi diagram for the points below?

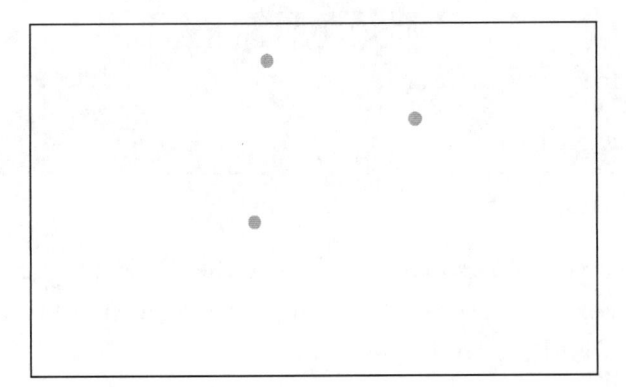

6. Brainstorm situations where Voronoi modeling would be helpful.

Protectors of the Realm

Off You Go

Activity 1: Castles Map

Goals:	To learn about your castle and locate it on a map; to create a Castles Map that shows all castle locations
Materials:	computer with Internet access, map of England, graph paper, pencils, ruler
Tools:	Medieval Castles in England and Wales

Directions

1. Follow your teacher's instructions for team formation. Find out the name of your castle. *Note:* Your castle is an actual medieval castle located in what is now England or Wales.

2. Using your Helpful Web Resources, research your castle.

3. On a separate sheet of paper, note the castle's history, who built it, why it was built, what it looks like, and so forth. Find out if it is still standing or in use today. Try using Google Maps or Google Earth to view a satellite image of your castle and/or the area.

4. Determine the location of your castle on a map of England. Record the longitude and latitude of the location. Note: If you can't learn the precise location of the castle, use the location of the village or town it was in.

5. Share and compare information on your castle with other teams. Take notes! Record the longitude and latitude coordinates and other location information for all castles.

6. Get a map of England for your team from your teacher.

7. As a team, mark the precise location of each castle on the map. Be sure to represent locations accurately (at scale). Your finished work is called your Castles Map.

Protectors of the Realm

Expedition Tool

Medieval Castles in England and Wales

Castle	Longitude	Latitude	Notes/features
Newark			
Stafford			
Ludlow			
Warwick			
Richmond			
Ruthin			
Wallingford			
Castle Rising			
Builth			
Berkeley			

Protectors of the Realm

Off You Go

Activity 2: The Extent of Your Domain

> **Goal:** To create a Voronoi diagram to define the boundaries of each lord's lands
>
> **Materials:** graph paper, pencils, ruler

Directions

1. Get together with your group. Get out your Castles Map created during the last activity.

2. Listen as your teacher provides information. Use the information to mark the outer limits of what will become your diagram.

3. Create Voronoi polygons that show the territorial boundaries or district of each castle.

4. Once complete, compare your diagram with those of other teams. Discuss your observations. Then answer the questions below.

   a. How do castle territories compare in size and shape?

   _____
   _____

   b. What do you notice about where castles are located relative to boundaries and other neighboring castles?

   _____
   _____

   c. Who has the most land? The least?

   _____
   _____

   d. Who has portions of coastline and sea?

   _____
   _____

   e. What are the apparent strengths of each territory? What are potential vulnerabilities and challenges?

   _____
   _____

Protectors of the Realm

Off You Go

Activity 3: Wise Counsel

> **Goal:** To use triangulation, other geometry skills, the Internet, and your imagination to recommend territory improvements to your lord
>
> **Materials:** graph paper, pencils, ruler

The lord of your territory is seeking your advice on the following:

a. If he were to build a new castle in a central location, where should it be?

b. He wants to hold an annual tournament that will involve three neighboring lords. He wants it in a central location. Where should the tournament be held?

c. The king has asked that he and two neighboring lords construct a large cathedral in a central location. Where should the cathedral be built?

d. Should he build any outposts or villages? For example, is there a key location to defend or an ideal location for a village market?

The lord wants your recommendations as soon as possible.

Final Recommendation Criteria

1. Include a scale map of England and Wales showing all castles and boundaries.
   - Label all castles.
   - Highlight your castle and its territory using colored pencils or another method.
   - Define the distance between castles as "as the crow flies."

2. Include a map that provides an enlarged view of your territory and any territories that lie alongside you.
   - Include the locations of castles and boundaries.
   - Mark the dimensions of your territory: the total area, length, width, height, and distance from the castle to each boundary line (choose at least two points).
   - Show and label any important geographic features (rivers, large lakes, oceans, mountains, and so forth).

(continued)

Protectors of the Realm

Off You Go

> 3. Include a written report that states and explains each of your recommendations.
>   - Discuss pros or cons of the location.
>   - Propose locations for a cathedral, the tournament, and a new castle.
>   - Propose locations for one or two additional construction projects to improve the territory.
>   - Show the geometry and calculations you used to make recommendations on the appropriate map(s).
>   - Include the distances from the castle(s) to the location.
>   - Include any other creative observations or recommendations you have.

Directions

1. Use your web resources or an atlas to find out more about the physical features of your land.

2. Create your scale maps.

3. Solve the three specific location problems (propose locations for a cathedral, the tournament, and a new castle). Show your work on your map.

4. Should the lord build anything else? If so, what, where, and why? Consider the shape and size of the territory, the location of the castle or neighboring castles, or the physical features of the area. Identify one or two additional construction projects. *Tip:* Make copies of your map and use the copies to work out location calculations. Show final work on your original map.

5. Prepare a draft of your recommendations. All team members should review it and make any necessary revisions.

6. As a team, decide how you will present your recommendations. Each team will have five minutes. Each team member should be involved in the presentation.

Protectors of the Realm

Check Yourself!

Skill Check

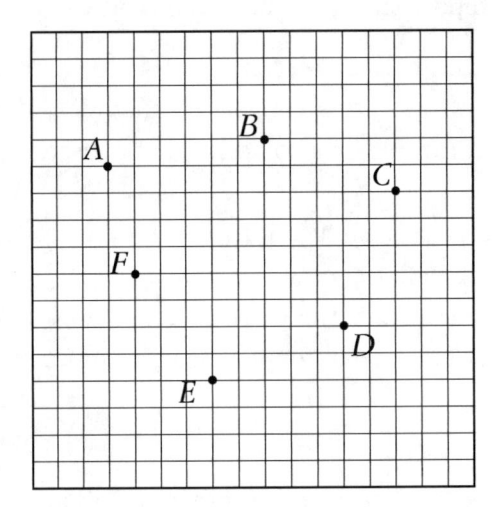

The grid above represents an area of the Serengeti National Park. Points A and E are water holes. The other points are lion prides. Use the grid to answer the questions below.

1. Which lion pride is closest to water hole A?

2. Is pride C closer to water hole A or E?

3. What area might each pride consider its territory? Show your answer on the grid.

4. Which prides have territory with water holes?

5. Which pride has the biggest territory? If each grid square represents $\frac{1}{4}$ square mile, how big is this area?

6. Where might park rangers create a new water source in order to minimize competition at the other water holes?

Protectors of the Realm

Check Yourself!

Self-Assessment and Reflection
Project Management

Before You Go

- ❏ I understand how to create a Voronoi diagram and how it is applied to the concept of "zone of proximity."
- ❏ I'm honestly not sure I understand the math involved in the project and have asked my teacher for additional help.

Off You Go

- ❏ I reviewed the activities and materials for this project and understood the products I had to create.
- ❏ I helped my team research our castle and its location, including longitude and latitude coordinates. I was prepared to relate location information and any interesting castle facts in class.
- ❏ I helped my team create our Castles Map and Voronoi diagram. We split the work evenly. Everyone did a portion of the map.
- ❏ I participated fully in class discussions to review our map and diagram.
- ❏ I used an atlas and other tools to learn more about the topography of our castle territory.
- ❏ I made significant, high-quality contributions to help our team create scale maps, identify good locations for events and buildings, and prepare our final recommendations report.
- ❏ I am prepared to speak about any aspect of our maps, recommendations, and report. I participated in our short presentation.

Do You Know?

- ❏ I can define the Lingo to Learn vocabulary terms for this project and give an example of each.
- ❏ I completed the Skill Check problems and carefully reviewed problems I answered incorrectly.

Protectors of the Realm

Check Yourself!

Reflection

1. What were the most challenging aspects of this project for you and why?

2. What skills did this project help you develop?

3. If you did this project again, what might you do differently and why?

Superhero Challenge

Overview
Student superheroes model geometric networks and apply graph theory using coordinate systems to figure out the shortest route they can take to respond to simultaneous distress calls from citizens of Schoolopolis.

This project is a variation of the classic "traveling salesperson" or "shortest path" problem using knowledge of coordinate systems and coordinate geometry.

Time
Total time: 4 to 7 hours

- Before You Go—Express Delivery: 20 to 30 minutes in class
- Activity 1—Schoolopolis City Grid: two to three 55-minute class periods
- Activity 2—Superhero Rescue Strategy: one to two 55-minute class periods or homework assignments

Skill Focus
- coordinate systems
- measurement
- geometric networks and network optimization

Prior Knowledge
- graphing Cartesian coordinates
- calculations involving angles and triangles
- measurement

Team Formation
Students work in teams of three or four students.

Lingo to Learn—Terms to Know
- **coordinate:** a set of numbers that determines the location of a point in space
- **edge:** the side of a polygon or polyhedron
- **graph theory:** mathematical study of the properties of graphs
- **Hamiltonian circuit:** a cycle through a graph that visits each node exactly once
- **node:** vertex; connection point
- **optimization:** search for the best solution among alternatives
- **vertex:** corner; a point where lines, rays, or polygon sides or polyhedron edges meet

Superhero Challenge

Suggested Steps

Preparation

- Determine how students will be divided into teams. It is important to consider because you should not add rescue sites if there are more than five teams. Teams should then combine forces to develop the five superhero rescue call scenarios.
- Choose five locations for Hero Hall (your classroom) and the four superhero rescue points. All locations should be on the same floor of the school (on the same plane). Small steps, inclines, or ramps are fine.
- Gain permission for students to access corridors and locations for Activity 1: Schoolopolis City Grid.
- Students will need a coordinate for each location. Identify a specific spot in each room or location. Put a sticker or draw a star on an index card or a sticky note. You will place the card or sticky note at the precise point you want students to use.
- If possible, get a copy of the school's floor plan with building and room dimensions. Use it only to confirm student work, to help them with measurements, or to create a simplified view.
- If providing students with a scale map (guided challenge), leave off coordinates and other measures they can calculate.

Day 1

1. Provide an overview of the project and review materials.

2. Ask students to give examples of situations that sound like shortest-path problems. Scenarios could involve a pizza-delivery person, a bike messenger, a 911 operator, or a tourist or travel agent planning a trip.

3. As a class, work on Before You Go: Express Delivery. Explain the following:
   - The location point is the *vertex*. The path between points is the *edge*. The vertex is sometimes called the *node*.
   - The "traveling salesperson" scenario is the classic example of the problem. A salesperson must decide the shortest or least expensive route to visit each customer. This is why it is often called the Traveling Salesperson Problem (TSP).
   - The assumption is that you can visit each vertex exactly once (except the start/end point). *Optional:* This is known as a Hamiltonian path or cycle.

4. Ask students to imagine the effort it would take to manually solve a large-scale TSP problem. Share the fact that for 10 points, it could take 3,628,800 calculations.

Expeditions in Your Classroom: Geometry   © Walch Education

Superhero Challenge

5. Assign teams. Give teams 5 or 6 minutes to choose superpowers. Alternatively, you can assign powers.

6. Provide each team with a rescue location. Tell students to brainstorm for about 5 minutes. This time, direct them to create and summarize a scenario on their Rescue Calls sheet. If you have more than four teams, assign the same location to several teams. Have students brainstorm and then vote to choose the scenario.

7. Call upon teams to briefly describe the scenario they devised.

Days 2 and 3

1. Call together the teams and review Activity 1: Schoolopolis City Grid.

2. Facilitate the creation of one grid map all teams can use. Decide whether the map will be drawn by hand or generated by computer. Ask students to organize to accomplish the task or ask for several volunteers.

3. Manage grid design as an open-ended or guided challenge.

Open-Ended Challenge

Simply state the problem: You have placed cards at five locations in the building. Each card represents a rescue point coordinate in Schoolopolis. The class must create a single grid with coordinates for Hero Hall and each rescue point. They must also show distances between points. Students can assume a direct path between points (a straight line regardless of walls and other obstacles). They may also treat the school as a single plane. Distance measurements do not need to include small rises or steps.

Ask students to strategize the best approach. Once students agree on a strategy, let them test it and revise as needed. Time needed for this approach will range from one to three class periods.

Suggestions might include the following:

- Manually measure distances between points and obtain other measurements needed to create a school floor plan. Use efficient measurement strategies and tools (for instance, a tape measure, a student's stride, a long rope of predetermined length, and so forth).
- Use a combination of measurement, estimation, and scaling strategy to create the grid. Define the rescue zone—only that area of the building that includes points involved. Measure only what is absolutely needed (for example, the length

Expeditions in Your Classroom: Geometry © Walch Education

Superhero Challenge

and width of the rescue zone, several rooms to see if there is a pattern to the dimensions, the distance between the rescue zone boundaries and a rescue point). Calculate point distances using coordinate geometry. Check a few measurements manually to compare results.

- Use GIS software to develop a grid of the school. Take only the measurements needed to identify point locations.
- Create the grid with known information and use coordinate geometry to calculate distances.
- Decide that creating a scaled grid of the school isn't important—one just needs a grid with values, points, and distances that everyone accepts. This is entirely true, at least as far as the superhero rescue problem itself goes. Discuss if creating a model versus a scale grid is enough practice with coordinate geometry and measurement.

Guided Challenge

Provide a scaled floor plan of the school with basic dimensions and a legend. Invent the dimensions, base them on the school's actual plan, or give students a rudimentary version of the actual plan. Have students use the information to pinpoint rescue locations and calculate distances. Students will still need to visit rescue site locations to collect measurements needed to plot coordinates.

4. Review the completed Schoolopolis City Grid as a class. Make one copy per team.

Day 4

1. Give each team a Schoolopolis City Grid.
2. Review Activity 2: Superhero Rescue Strategy.
3. Have students start work in class and continue as homework.

Homework

Have students finish Activity 2: Superhero Rescue Strategy.

Superhero Challenge

Day 5

1. Review team findings. Ask teams which three paths they recommend. What is each team's preferred path and why? Note that reasons may relate more to superpowers and obstacles than to mathematical analysis.

2. If time allows, discuss variations, applications, or more complex situations:
   - Would it have made a difference if students had selected a different starting point?
   - Now that students understand the "traveling salesperson" problem, what applications do they see? Who else would face this dilemma on a day-to-day basis (the airline industry, parcel pickup companies, emergency management)?

Final Day

1. Have students complete the Skill Check problems.
2. Check and review answers.
3. Have students complete the Self-Assessment and Reflection worksheet and submit it (optional).

Project Management Tips and Notes

- Each additional point makes this problem far more complex. To simplify the problem, use only four rescue sites and Hero Hall (five points total). Using more than five points is not recommended.
- There are several ways you can manage Activity 1: Schoolopolis City Grid. See step 3, Days 2 and 3. Decide on the approach you want to use before you begin the project. Prepare a school floor plan in advance.

Suggested Assessment

Use the Geometry Project Assessment Rubric or the following point system:

Team participation	25 points
Schoolopolis city grid/class participation	30 points
Rescue strategy	40 points
Project self-assessment	5 points

Expeditions in Your Classroom: Geometry © Walch Education

Superhero Challenge

Extension Activities

- For added amusement, have students create a picture of their superhero alter ego. Alternatively, allow them to dress up as the superheroes.
- Challenge students to solve additional shortest-path problems based on other applications: an ant colony where ants find optimal paths to food, a parcel pickup company, an emergency rescue squad, or a traveling salesperson (where cost and distance might figure into the analysis).
- Have students use GIS software to help solve the problem.
- Explain algorithms and show students how an algorithm could be created to solve similar problems.
- Explore "taxicab geometry," in which paths follow city blocks rather than straight lines.

Other Helpful Resources

- Georgia Institute of Technology—The Traveling Salesman Problem
  www.tsp.gatech.edu/index.html

- Wolfram MathWorld—Traveling Salesman Problem
  http://mathworld.wolfram.com/TravelingSalesmanProblem.html

Common Core State Standards Connection

High School
Number and Quantity: Quantities★

N-Q.1. Use units as a way to understand problems and to guide the solution of multi-step problems; choose and interpret units consistently in formulas; choose and interpret the scale and the origin in graphs and data displays.

Grade 8
Geometry

8.G.8. Apply the Pythagorean Theorem to find the distance between two points in a coordinate system.

Expeditions in Your Classroom: Geometry

Superhero Challenge

Answer Key
Before You Go: Express Delivery

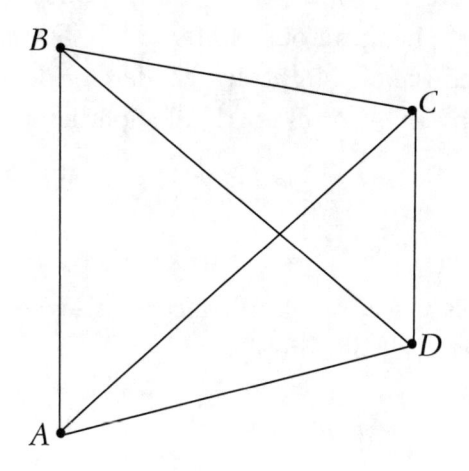

Path $A > B > C > D > A$ is shortest: 12.7 miles.

Path $A > C > D > B > A$ is next best: 14.9 miles.

The other paths are 15-mile journeys.

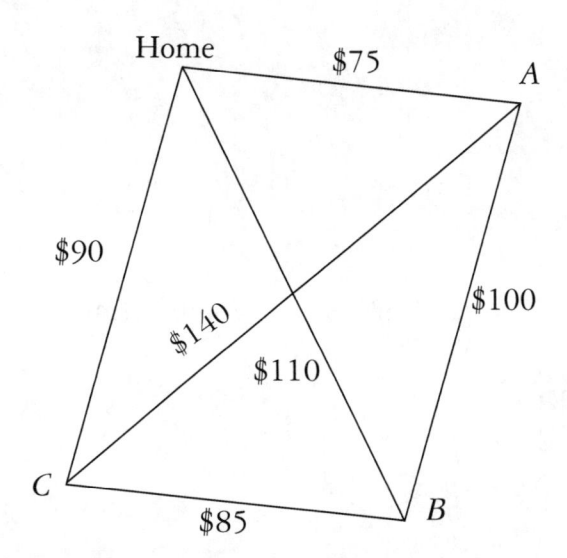

Check Yourself! Skill Check

Least expensive trip: Home > City A > City B > City C > Home ($350)

Most expensive trip: Home > City B > City A > City C > Home ($440)

Superhero Challenge

Expedition Overview

Challenge
You are a member of a team of superheroes. Each of you has an extraordinary power that, when combined with other superheroes' powers, can get any citizen of Schoolopolis out of trouble. You and your colleagues are hanging out at Hero Hall. Suddenly, distress signals from four different locations around the city light up the alert grid. As a team, you need to respond to all of the calls by finding the shortest path all superheroes can take to reach each location and save everyone.

Objectives
- To model geometric networks and apply graph theory using coordinate systems
- To solve "shortest-path" (optimization) problems

Project Activities
Before You Go
- Express Delivery

Off You Go
- Activity 1: Schoolopolis City Grid
- Activity 2: Superhero Rescue Strategy

Expedition Tool
Rescue Calls

Other Materials Needed
- graph paper
- pencils
- ruler
- tape measure (the longer, the better), yardstick, or roll of string
- calculator
- superhero costume (optional)

Expeditions in Your Classroom: Geometry

Superhero Challenge

Expedition Overview

Lingo to Learn—Terms to Know
- coordinate
- edge
- graph theory
- Hamiltonian circuit
- node
- optimization
- vertex

Helpful Web Resources
- Georgia Institute of Technology—The Traveling Salesman Problem
  www.tsp.gatech.edu/index.html

- Wolfram MathWorld—Traveling Salesman Problem
  http://mathworld.wolfram.com/TravelingSalesmanProblem.html

Superhero Challenge

Before You Go

Express Delivery

> **Goal:** To understand shortest-path problems
>
> **Materials:** pencils, ruler

Directions

You are a busy pizza delivery person with many pies to deliver before you call it quits for the night. Three orders are up and ready for customers. What is the best way for you to get to all delivery locations in the least amount of time?

Create a graph to model your situation. Assume each grid square is 1 mile. Identify the shortest route you can take from the pizza restaurant to all customer houses and back again.

Location	Coordinates
Pizza place (A)	3, 3
Customer 1 (B)	4, 7
Customer 2 (C)	7, 6
Customer 3 (D)	6, 4

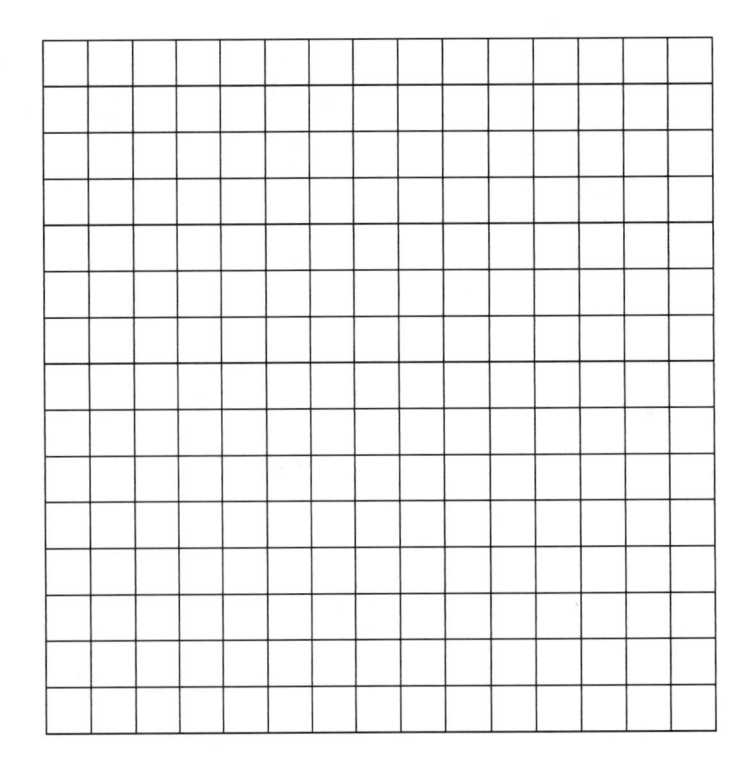

Superhero Challenge

Off You Go

Activity 1: Schoolopolis City Grid

Goals:	To develop your superhero identity and review rescue call situations; to create a grid map of your school that provides the coordinates of Hero Hall and rescue locations
Materials:	graph paper, pencils, ruler
Tools:	Rescue Calls

Directions

1. Meet with your team and review the Rescue Calls information sheet.

2. Brainstorm your superpowers and superhero identities. Remember, your team must work together to resolve each crisis. You cannot dispatch different members to different locations.

3. Your teacher will provide rescue location information. Record any information you receive on your Rescue Calls sheet.

4. Brainstorm a rescue scenario for one of the locations. Describe the scene on the Rescue Calls sheet. Use your imagination!

5. As a class, create a Schoolopolis City grid map. Your grid must show Hero Hall and each rescue site as coordinate points and distances between points. Follow your teacher's instructions for this challenge.

Schoolopolis City Grid Criteria

- ❏ The Schoolopolis grid does not need to be elaborate. Include enough of the school layout, floor plan, or reference points so that rescue site locations and the paths to them are recognizable.

- ❏ Indicate the scale of your grid and units of measure.

- ❏ Label all points. Provide coordinates.

- ❏ Indicate the distance between all points. Assume you can travel directly from one point to another, even if this means walking through walls. After all, you are superheroes!

- ❏ Include Hero Hall, your starting point.

Expeditions in Your Classroom: Geometry © Walch Education

Superhero Challenge

Expedition Tool

Rescue Calls

Record the location information your teacher provides. Next, brainstorm an interesting emergency situation. Describe your situation below. Think about the location and the type of scenario that would benefit from superhero solutions. Be imaginative!

Hero Hall   Coordinates: _____

Rescue location 1: _____   Coordinates: _____

_____
_____
_____
_____
_____
_____
_____
_____

Rescue location 2: _____   Coordinates: _____

_____
_____
_____
_____
_____
_____
_____
_____

(continued)

Superhero Challenge

Expedition Tool

Rescue location 3: _____   Coordinates: _____

_____
_____
_____
_____
_____
_____
_____

Rescue location 4: _____   Coordinates: _____

_____
_____
_____
_____
_____
_____
_____

Superhero Challenge

Off You Go

Activity 2: Superhero Rescue Strategy

> **Goal:** To identify and rank the shortest paths you can take to reach all rescue points
>
> **Materials:** graph paper, colored pencils, ruler, protractor

Directions

1. Using your Schoolopolis City Grid and distance measurements, identify and rank the three shortest paths your team could take to rescue everyone. List your best paths below.

   Remember:

   - You must start at Hero Hall.
   - You must visit each rescue point together.
   - You must travel in a direct line between points.
   - You may only visit each point once. You cannot revisit it en route to another point.

   Shortest Paths

Rank	Rescue order (location and name)	Total path distance
1		
2		
3		

2. Illustrate the three shortest paths on graph paper. Create multiple graphs or use colored pencils to differentiate options on the same grid.

3. Decide which path is ultimately best for your team. Explain your rationale (obstacles, your superpowers, and so forth).

   _____
   _____
   _____
   _____
   _____

Superhero Challenge

Check Yourself!

Skill Check

It's election time. A gubernatorial candidate needs to hit the campaign trail. She knows that voters in three cities can make or break the race for her.

Given the costs of flying from city to city, what is the least expensive way for her to visit all three cities in one trip and return home? What is the most expensive route?

Route	Airfare
Home to City A	$75
Home to City B	$110
Home to City C	$90
City A to City B	$100
City A to City C	$140
City B to City C	$85

Illustrate your answer using the graph below.

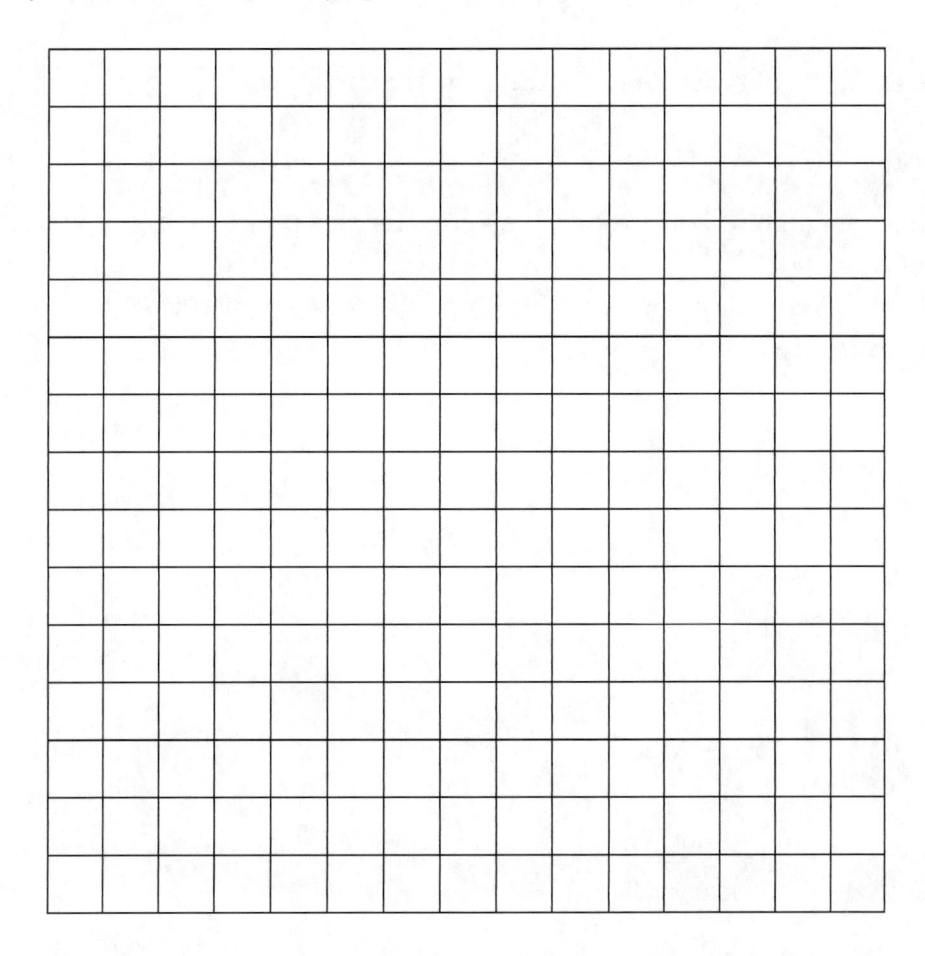

Superhero Challenge

Check Yourself!

Self-Assessment and Reflection
Project Management

Before You Go

- ❏ I understand the concept of a geometric network and how to use a graph to represent one.
- ❏ I know how to calculate the distance between points on a graph.
- ❏ I'm honestly not sure I understand the math involved in the project and have asked my teacher for additional help.

Off You Go

- ❏ I reviewed the activities and materials for this project. I understood the problem I had to represent and solve.
- ❏ I helped my team develop superpower personalities and write a scenario for one rescue site.
- ❏ I contributed as our class developed a strategy for creating the Schoolopolis City grid.
- ❏ I collected or calculated measurements we needed for the grid.
- ❏ I helped my team figure out how to solve our shortest-path problem. I analyzed and calculated distances for specific routes.
- ❏ I can give examples of other shortest-path scenarios.

Do You Know?

- ❏ I can define the Lingo to Learn vocabulary terms for this project and give an example of each.
- ❏ I completed the Skill Check problems and carefully reviewed problems I answered incorrectly.

Superhero Challenge

Check Yourself!

Reflection

1. What were the most challenging aspects of this project for you and why?

   _____
   _____
   _____
   _____
   _____
   _____
   _____

2. What skills did this project help you develop?

   _____
   _____
   _____
   _____
   _____
   _____
   _____

3. If you did this project again, what might you do differently and why?

   _____
   _____
   _____
   _____
   _____
   _____
   _____

Thinking Outside the Box

Overview
Students design a new container for a product in an unconventional shape. They create scaled drawings of three different views of their proposed container, build a three-dimensional model, and tout the merits of their design in a one-page proposal.

Time
Total time: 7 to 10 hours

- Before You Go—Solid Engineering: one 55-minute class period
- Activity 1—Product Hunt: one to two 55-minute class periods and 30 minutes of homework
- Activity 2—Package Redesign: two to three 55-minute class periods and $2 \frac{1}{2}$ hours of homework
- Activity 3—Pitch the Executives: 45 to 60 minutes of homework and one 55-minute class period

Skill Focus
- solid geometry (spheres, pyramids, cones, prisms, Platonic solids)
- measurement and scale

Prior Knowledge
- classifying solids
- calculating solid dimensions
- perspective views of three-dimensional figures

Team Formation
Students work in groups of two or three students.

Lingo to Learn—Terms to Know
- **altitude:** height
- **base:** the bottom of a plane figure or a three-dimensional figure
- **face:** a flat surface of a three-dimensional figure
- **lateral edge:** an edge of a polyhedron; usually refers to a prism or a pyramid (prism: the intersection of two adjacent lateral faces; pyramid: the edges of the lateral faces that join the vertex to the vertices of the base)
- **Platonic solids:** convex polyhedrons with equivalent faces composed of congruent convex regular polygons (there are five: cube, dodecahedron, icosahedron, octahedron, and tetrahedron)

Thinking Outside the Box

- **rotation:** a transformation in which a figure is rotated through a given angle, about a given point
- **solid:** a three-dimensional figure
- **vertex:** corner; a point where lines, rays, or polygon sides or polyhedron edges meet

Suggested Steps

Preparation

- Review Activity 2: Package Redesign. If desired, support the focus of your unit by changing or narrowing the criteria.
- Obtain six or seven product containers (one per group) for Before You Go: Solid Engineering. Choose a variety of shapes and/or sizes, all basic geometric solids. If possible, choose cardboard containers that students can disassemble to examine construction. Otherwise, use what you can find. As a last resort, use a solid object instead of a container. Ask teachers and friends to keep an eye on their home recycling box for interesting prospects. Remember to cover package volume information with masking tape. Suggestions: triangular prism—Toblerone chocolate bar container; rectangular prism—plastic wrap container, tissue box, cereal box, cracker box, toiletry box; cone—Hershey's Kiss, cotton candy stick, ice cream cone wrapper; cylinder—oatmeal container, ice cream container, tissue box, lip balm, coffee cup, first aid tape (with a hole); sphere—shampoo bottle, perfume bottle.
- You might also find other interesting packaging examples—either the real thing or pictures. They don't have to be cardboard or perfect geometric solids. Ask students to tell you the basic shape(s) they use.

Day 1

1. Provide an overview of the project and review materials.
2. Explain the term container (product packaging) and show several examples. Ask students for other examples to make sure they understand the variations (a bottle within a box and so forth).
3. Facilitate Before You Go: Solid Engineering to review solids and their properties.
4. Explain Activity 1: Product Hunt. Assign steps 1–2 for homework. Review examples of items students are permitted (and not permitted) to bring in.

Homework

Have students find one or two examples of product packaging to bring to class.

Thinking Outside the Box

Teacher page

Day 2

1. Use the Packaging Analysis Worksheet to guide a discussion about the products and containers students brought in. Students may need one or two examples of "functions" (tamperproof features, handles for carrying, dust protection, and so forth).

2. Call on students to suggest strategies for calculating or estimating the dimensions of items. Alternatively, form small groups, give each group a container, and ask for dimensions.

3. Form teams. Direct them to identify the product container they want to redesign.

4. You may allow teams to choose any product they want. Alternatively, provide an opportunity for students to see varied approaches for the same item. There are two ways to do this:

   - As a class, vote on two or three interesting options. Have each team choose the option they want.

   - Predetermine a product. Ask all teams to submit a design for the same product.

5. If students will use an item already in the classroom, provide time for them to examine it and collect information. Otherwise, assign selecting and analyzing their product or container as homework.

Homework

Have students find a product or container for their redesign project.

Days 3 through 7

1. Explain Activity 2: Package Redesign. Indicate if work will be done in class, at home, or both. Assign due dates for the three sketches and the 3-D model.

2. Review Container Criteria. Add any other criteria you have. Direct students to focus on particular shapes.

3. Allow teams to work at their own pace. Circulate as students work. Pause every now and then to highlight good examples of technique and problem-solving effort.

4. Review each team's finished drawings before students continue to the model-building step. If organized as a class activity, provide poster board and materials needed.

5. Provide or assign time for students to complete drawings and model building.

Thinking Outside the Box

Day 8

1. Explain Activity 3: Pitch the Executives.

2. Review the Container Proposal Worksheet.

3. Allow teams to begin work on the proposal in class. Give a due date for the proposal and presentation.

4. Remind students to attach their scale drawings to their proposals.

Presentation Day

1. Allow each team five minutes to present.

2. Solicit feedback from other students and discuss.

3. Collect a container proposal and three scale drawings from each team.

Final Day

1. Have students complete the Skill Check problems.

2. Check and review answers.

3. Have students complete the Self-Assessment and Reflection worksheet and submit it (optional).

Project Management Tips and Notes

- While there are several options for identifying a product or products for Activity 2: Package Redesign, focusing everyone on the same product can also work well. You may see wildly different container conceptions or a convergence of design ideas. It can depend on the product. If you choose this option, you may want to issue an additional challenge: Ask teams to choose the name of a solid shape from a hat and incorporate that shape into their design, alone or in combination with another shape. If you ask teams to choose from a limited set of two or three products, they have a choice but can still see how different groups approach the same challenge.

Thinking Outside the Box

- Students may want clarification on how far they can take their design. Recommend that they keep it relatively simple, since they must build it.
- Try to do some drawing work in class. This is the most challenging step for many students. Note that the blueprint will give them what they need to create their 3-D model.

Suggested Assessment
Use the Geometry Project Assessment Rubric or the following point system:

Team and class participation	15 points
Three scale drawings/views	45 points
3-D container model	35 points
Project self-assessment	5 points

Extension Activities
- Expand upon the "efficiency" analysis. Ask students to calculate the shape and dimensions of the box needed to ship 12 units of the newly designed product. Also have students compare the amount of shelf space needed for 20 units of the old design and 20 units of the new design. Pull examples from student designs or use a container you provide.
- Assign students to create a digital model of their container using a computer-aided design (CAD) tool.
- Have students draw solids from a variety of perspectives. Set up a sample container in the middle of a table. Give students 2 to 4 minutes to quickly sketch it from their spot. Rotate the container and have students draw it again. Repeat with other solids.

Common Core State Standards Connection
High School

Geometry: Geometric Measuring and Dimension

G-GMD.3. Use volume formulas for cylinders, pyramids, cones, and spheres to solve problems.★

Geometry: Modeling with Geometry

G-MG.1. Use geometric shapes, their measures, and their properties to describe objects (e.g., modeling a tree trunk or a human torso as a cylinder).★

G-MG.3. Apply geometric methods to solve design problems (e.g., designing an object or structure to satisfy physical constraints or minimize cost; working with typographic grid systems based on ratios).★

Thinking Outside the Box

Number and Quantity: Quantities★

N-Q.1.   Use units as a way to understand problems and to guide the solution of multi-step problems; choose and interpret units consistently in formulas; choose and interpret the scale and the origin in graphs and data displays.

Grade 8
Geometry

8.G.9.   Know the formulas for the volumes of cones, cylinders, and spheres and use them to solve real-world and mathematical problems.

Grade 7
Geometry

7.G.1.   Solve problems involving scale drawings of geometric figures, including computing actual lengths and areas from a scale drawing and reproducing a scale drawing at a different scale.

Grade 6
Geometry

6.G.4.   Represent three-dimensional figures using nets made up of rectangles and triangles, and use the nets to find the surface area of these figures. Apply these techniques in the context of solving real-world and mathematical problems.

Answer Key
Check Yourself! Skill Check

1. a. 400,000 ft^3

   b. 561 in^3

   c. 135 cm^3

2. Design C, the rectangular prism, is the best choice. If the cat-food containers are packed right side up, you can fit 445 containers per shipping box for a total volume of 4,672.5 cubic inches. This is approximately 28% more containers and 23% more total volume than with design A, the second best option. The maximum number of cans you can fit with design A is 320 (total volume of 3,616 cubic inches) and with design B is 355 (total volume of 3,763 cubic inches).

Thinking Outside the Box

Expedition Overview

Challenge
You work for Creative Containers, a company that designs boxes and containers for retail products. Recently, MakeItAll Inc., an international consumer products business, turned to you for fresh ideas—and fresh shapes. They have asked you to propose new package designs for several existing products and present them to company executives.

Objectives
- To learn the properties of solids and how to calculate the dimensions of a solid figure
- To learn how to represent three-dimensional objects from different perspectives
- To understand how to create a blueprint for a three-dimensional object

Project Activities
Before You Go
- Solid Engineering

Off You Go
- Activity 1: Product Hunt
- Activity 2: Package Redesign
- Activity 3: Pitch the Executives

Expedition Tools
- Packaging Analysis Worksheet
- Container Proposal Worksheet

Other Materials Needed
- graph paper
- paper
- pencils
- colored pencils or crayons
- ruler
- poster board or lightweight cardboard
- glue or tape
- scissors
- art supplies (markers, paint, construction paper, scissors)

Thinking Outside the Box

Expedition Overview

Lingo to Learn—Terms to Know
- altitude
- base
- face
- lateral edge
- Platonic solids
- rotation
- solid
- vertex

Helpful Web Resources
- The American Package Museum
  www.packagemuseum.com

- Math.com—Three-Dimensional Figures
  www.math.com/school/subject3/lessons/S3U4L1GL.html#

- The Math Forum—3-D Drawing and Geometry
  http://mathforum.org/workshops/sum98/participants/sanders/Isom.html

- National Council of Teachers of Mathematics—Illuminations: Geometric Solids
  http://illuminations.nctm.org/ActivityDetail.aspx?ID=70

- *Package Design* Magazine—Spotlight
  www.packagedesignmag.com
  (See "Spotlight" section on the right-hand side of the page.)

- *Packaging World*
  www.packworld.com

- University of Surrey—School of Electronics and Physical Sciences: Some Solid (Three-Dimensional) Geometrical Facts About the Golden Section
  www.mcs.surrey.ac.uk/Personal/R.Knott/Fibonacci/phi3DGeom.html

Thinking Outside the Box

Before You Go

Solid Engineering

> **Goal:** To review the basic properties of solids
>
> **Materials:** container samples (provided by teacher), rulers, graph paper

Directions

1. Form a group with one or two other students. Your teacher will provide you with a container.

2. Calculate and record the dimensions of your container. Measurements needed may vary from container to container.

   Shape of container (be specific): _____

   Height: _____

   Altitude: _____

   Width: _____

   Depth: _____

   Volume: _____

   Total surface area: _____

   Face: _____

   Base: _____

   Vertex: _____

   Lateral edge: _____

   Circumference: _____

   Diameter: _____

3. Examine the construction of the container from the outside and the inside.

4. If you can safely disassemble your container, do so. Preserve any folds, flaps, or seams that hold it together.

(continued)

Thinking Outside the Box

Before You Go

5. What do you notice about the construction of the container? Write your observations below.

   _____
   _____
   _____
   _____
   _____

6. Sketch a blueprint of your disassembled container below or on a separate sheet of paper. If you can't take the container apart, draw what you think it would look like.

Thinking Outside the Box

Off You Go

Activity 1: Product Hunt

Goals:	To analyze examples of product containers and packaging; to identify products for the container redesign competition
Materials:	paper, ruler
Tools:	Packaging Analysis Worksheet

Directions

1. Think about boxes and containers that you see at home or in a store. In the space below, make a list of products that come in interesting packages.

2. Look at your list. Note the shapes used for packages. Which shapes are used most often?
   _____
   _____
   _____
   _____

3. Are there any patterns you notice?
   _____
   _____
   _____
   _____

(continued)

Thinking Outside the Box

Off You Go

4. What relationship is there between a product and its container type or shape?

   _____
   _____
   _____
   _____

5. For homework, find one or two items that you would like to redesign a package for. Use the following guidelines:

   - Your container can be any shape.
   - It should be small enough to carry.
   - Your item can be the packaging a product comes in or the container that actually holds the product.
   - Your container should be empty, unless it holds materials normally allowed in school (for instance, a juice box, a CD or DVD case, and so forth). Check with your teacher if you are unsure.
   - You may bring an item that is in need of a container (for example, a soccer ball), whether or not the item had original packaging to begin with.

6. Before bringing in the item, cover up any volume information using tape, paper, or other method.

7. As a class, discuss the products and containers you brought in. Use the Packaging Analysis Worksheet.

8. Follow your teacher's instructions to get your redesign assignment.

9. With your team members, look at your product container carefully. Use the Packaging Analysis Worksheet to evaluate your container.

Thinking Outside the Box

Expedition Tool

Packaging Analysis Worksheet

Use this worksheet to help you analyze and discuss sample containers.

Appearance

Which containers are visually appealing? Which have a look that seems especially well-suited to the product they contain? Are visual elements important?

_____
_____
_____
_____
_____

Function

Which containers seem designed for strength? What makes a container strong? What other purposes or functions does the packaging have?

_____
_____
_____
_____
_____

Efficiency

Container production, shipping, and storage costs are important considerations when designing product packaging. Which containers seem like they might be more or less expensive to produce? Which might require additional packaging for safe shipping?

_____
_____
_____
_____
_____

(continued)

Thinking Outside the Box

Expedition Tool

Container Analysis

Product name: _____

Manufacturer: _____

Weight and volume information: _____

Other package features (product description, logo, bar code, health information or ingredients, and so forth):

_____
_____
_____
_____

Thinking Outside the Box

Off You Go

Activity 2: Package Redesign

Goal:	To generate a redesign plan and create scaled drawings to illustrate your plan
Materials:	paper, graph paper, ruler, poster board or cardboard, tape, glue, scissors, art supplies

Directions

1. As a team, review the criteria for your new container. Brainstorm redesign ideas!

   Container Criteria

   ❏ Your new container must have a different shape. You may use more than one shape.

   ❏ You cannot use only a cube or a rectangular prism. You may use these shapes in combination with other shapes.

   ❏ Your new design must hold the same volume as the current container.

   Add any other criteria your teacher provides:

   _____

   _____

2. Decide on a design idea.

3. Create three drawings that illustrate your design proposal:

   a. the container face (a view of the front of your container)

   b. a 3-D view (a drawing that shows a three-dimensional perspective of your container)

   c. the container template (an "unfolded" view or blueprint of your container that includes the flaps or folds needed to assemble it)

   Each drawing should be done to scale and should include the appropriate container measurements (metric). You may also include artwork and other details that show the "new look" of the product.

4. Use lightweight cardboard or poster board to create a three-dimensional model of your container.

5. Decorate your model to show the new look. Use art supplies or materials you print from your computer (such as the product logo). Be sure to include the product name and other visual elements normally found on packaging.

Expeditions in Your Classroom: Geometry

Thinking Outside the Box

Off You Go

Activity 3: Pitch the Executives

> **Goal:** To prepare and present your container design proposal
>
> **Tools:** Container Proposal Worksheet

Get ready to present your design idea to company executives!

Directions

1. As a team, use the Container Proposal Worksheet to prepare a brief, one-page proposal that summarizes your redesign idea. Explain why your design is effective. Attach your scale drawings to your proposal.

2. Decide how your team will present the proposal. Consider the following:

   - You will have five minutes to present your proposal.
   - Each team member must play a role.

3. Make your pitch—and be ready for questions!

Thinking Outside the Box

Expedition Tool

Container Proposal Worksheet

Your one-page proposal should include the following information and sections.

Date: _____

Team name: _____

Team members: _____

Product name: _____

Product description (what the product is, who it is for):
_____
_____
_____

Proposed design (two or three sentences describing your idea):
_____
_____
_____
_____
_____

Why customers will love it:
_____
_____
_____
_____
_____
_____

(continued)

Thinking Outside the Box

Expedition Tool

New container specifications (shape, dimensions, volume):
_____
_____
_____
_____

Efficiency (how the new design affects the amount of material needed to produce or ship the product):
_____
_____
_____
_____
_____
_____

Other comments:
_____
_____
_____
_____
_____

Attach your three design drawings to your proposal.

Thinking Outside the Box

Check Yourself!

Skill Check

1. What is the volume of each solid?

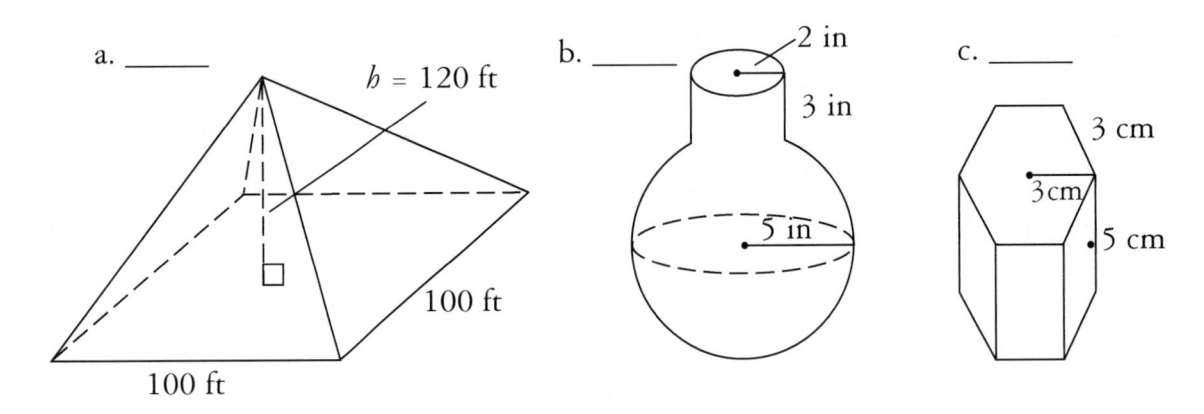

   a. _____ $h = 120$ ft, 100 ft, 100 ft

   b. _____   2 in, 3 in, 5 in

   c. _____   3 cm, 3 cm, 5 cm

2. Yummy Pet Food Company is weighing the pros and cons of three new cat-food container designs:

   Design A: a cylinder with a radius of 1.25 inches and a height of 2.25 inches

   Design B: a cylinder with a radius of 1.5 inches and a height of 1.5 inches

   Design C: a rectangular prism with a length of 4 inches, a width of 1.5 inches, and a height of 1.75 inches

   An important consideration is shipping. The company's standard shipping box has a length of 20 inches, a width of 20 inches, and a height of 12 inches. Which one is the best design for Yummy Pet Food Company? Why?

Thinking Outside the Box

Check Yourself!

Self-Assessment and Reflection
Project Management

Before You Go

- ❏ I can identify a variety of solid figures, describe their properties, and calculate dimensions.
- ❏ I'm honestly not sure I understand the math involved in the project and have asked my teacher for additional help.

Off You Go

- ❏ I reviewed the activities and materials for this project and understood the products I had to create.
- ❏ I scoped out interesting packages at home or at a local store. I brought one or two items to class as possible candidates for our container redesign project.
- ❏ I participated actively in our class analysis of container samples. I can give examples of container shapes, features, and reasons manufacturers might prefer one shape or design over another.
- ❏ I helped my team inspect the product and container we redesigned. I took notes on important information.
- ❏ I reviewed and understood the requirements of our new container (Container Criteria).
- ❏ I made contributions to the design of our container.
- ❏ I helped my team divide design-drawing tasks evenly and completed all of my tasks on time. We worked together to produce three high-quality scale drawings of our design.
- ❏ I helped my team create the three-dimensional model of our container. We shared the work evenly and completed the assignment by the due date.
- ❏ I contributed to our Container Proposal. I checked to make sure the proposal includes all required information. I am prepared to discuss any part of it and our scale drawings.
- ❏ I helped my team plan our presentation and played an active role delivering it.

Do You Know?

- ❏ I can define the Lingo to Learn vocabulary terms for this project and give an example of each.
- ❏ I completed the Skill Check problems and carefully reviewed problems I answered incorrectly.

Thinking Outside the Box

Check Yourself!

Reflection

1. What were the most challenging aspects of this project for you and why?

2. What skills did this project help you develop?

3. If you did this project again, what might you do differently and why?

Director's View

Overview
Students create a set map and solve field-of-vision problems based on a favorite movie scene. The project uses camera views and the art of cinematography to help students practice solving distance and measurement problems using the attributes of circles, angles, and triangles.

Time
Total time: 5 to 6 hours

- Before You Go—Camera Geometry—Field of Vision: 30 to 55 minutes
- Before You Go—Cinematography 101: 30 to 45 minutes
- Activity 1—Scene Analysis: one to two 55-minute class periods or homework assignments
- Activity 2—Set Map: one to two 55-minute class periods or homework assignments

Skill Focus
- circle and triangle geometry (tangent)
- scale representations

Prior Knowledge
- angle measurements
- basic properties of circles, angles, and triangles
- Some knowledge of floor plans and drawing to scale is also helpful.

Team Formation
Students work individually or in pairs.

Lingo to Learn—Terms to Know
- **arc:** a portion of the circumference of a circle
- **central angle:** an angle whose vertex is the center of the circle and whose sides pass through a pair of points on the circle
- **chord:** a line segment that joins two points on a circle
- **circle:** the set of points on a plane at a certain distance (radius) from a certain point (center)
- **field of vision:** all of the points that can be seen at a given moment without shifting gaze

Director's View

- **inscribed angle:** an angle placed inside a circle with its vertex on the circle and whose sides contain chords of the circle
- **major arc:** an arc with endpoints that form an angle more than 180 degrees with the center of the circle
- **minor arc:** an arc with endpoints that form an angle less than 180 degrees with the center of the circle
- **perpendicular:** forming a right triangle; meeting at a 90-degree angle

Suggested Steps

Preparation

- Choose a movie scene—one with interesting camera work—that you can use to help introduce the project. Films by Orson Welles or Alfred Hitchcock are always good choices.
- If possible, borrow a video camera and tripod you can use to help students understand field of vision.

Day 1

1. Give an overview of the project and review project materials.
2. Show a favorite film scene—twice. The second time, ask students to focus on the camera action (have students imagine that they are the camera). Discuss their observations.
3. Assign Before You Go: Camera Geometry—Field of Vision. If you have access to a video camera, ask students to experiment with "filling the picture."
4. Review Before You Go: Cinematography 101 and assign as homework.

Homework

Have students complete Before You Go: Cinematography 101.

Day 2

1. Facilitate Before You Go: Camera Geometry—Field of Vision. Work with students to help them understand the field-of-vision geometry involved in this project. Note: You may want to provide introductory information or problems if students need a refresher on circle and angle geometry.
2. Review Activity 1: Scene Analysis.

Expeditions in Your Classroom: Geometry

Director's View

3. Assign numbers 1–6 as homework. Tell students they must have at least two distinct shots (from the same camera or two). They may have up to five shots.

4. Number 7 is optional.

Homework (1–2 nights)

- Have students identify their scene, print the script for it, and analyze camera action.
- Have students list the shots they would use to (re)shoot the scene on the Shot Sheet.

Day 3

1. Introduce and explain the product and requirements of Activity 2: Set Map. Remind students that they must use at least two cameras. Both cameras must move at some point during the shoot.

2. Discuss the field-of-vision problems involved.

3. Explain that the maximum range of motion is the maximum distance the camera would travel in all directions for all shots one wants for a scene. Remind students that they must show their work.

4. Assign the due date for the map. Specify if class time will be used or if students must complete the assignment as homework.

Set Map Due Date

1. Call upon students to display their maps, describe their scenes, and give examples of camera location and motion.

2. Discuss what was challenging for students, how students solved problems, how students' paper cameras compared with real cameras, whether or not cinematographers are so calculated about shots, and so forth.

3. Collect final maps.

Final Day

1. Have students complete the Skill Check problems.

2. Check and review answers.

3. Have students complete the Self-Assessment and Reflection worksheet and submit it (optional).

Director's View

Project Management Tips and Notes
Tell students that it is not critical to draw the set maps precisely to scale. Students' measurements, however, should be precise. Students can do their field-of-vision calculations on graph paper and transfer measurements to the map.

Suggested Assessment
Use the Geometry Project Assessment Rubric or the following point system:

Team and class participation	15 points
Cinematography research	15 points
Script review/shot list	20 points
Set map	45 points
Project self-assessment	5 points

Extension Activities
- Assign students to use their measurements to create a 3-D model of their sets.
- As a class, explore the history of cameras, how cameras work, and the intersection of geometry and optics. Compare traditional and digital cameras. Investigate the future of cameras (for instance, the ability to film in 3-D).
- Have students investigate the geometry of 3-D animation and special-effects artistry.

Common Core State Standards Connection
High School

Geometry: Modeling with Geometry

G-MG.1. Use geometric shapes, their measures, and their properties to describe objects (e.g., modeling a tree trunk or a human torso as a cylinder).★

G-MG.3. Apply geometric methods to solve design problems (e.g., designing an object or structure to satisfy physical constraints or minimize cost; working with typographic grid systems based on ratios).★

Geometry: Similarity, Right Triangles, and Trigonometry

G-SRT.6. Understand that by similarity, side ratios in right triangles are properties of the angles in the triangle, leading to definitions of trigonometric ratios for acute angles.

Number and Quantity: Quantities★

N-Q.1. Use units as a way to understand problems and to guide the solution of multi-step problems; choose and interpret units consistently in formulas; choose and interpret the scale and the origin in graphs and data displays.

Director's View

N-Q.3. Choose a level of accuracy appropriate to limitations on measurement when reporting quantities.

Grade 7
Geometry

7.G.1. Solve problems involving scale drawings of geometric figures, including computing actual lengths and areas from a scale drawing and reproducing a scale drawing at a different scale.

Answer Key

Before You Go: Camera Geometry—Field of Vision

4. $\tan 29° = 87.5/d$
   $d = 87.5/\tan 29°$
   $d = 87.5/0.5543$
   $d = 158$ feet

Before You Go: Cinematography 101

1. distance, angle, and movement
2. A long shot: to establish the scene or view; to show surrounding action, context, or environment; to make individuals less of a focus. A close-up: to show or suggest emotion, tension, or guilt.
3. An eye-level shot: to show directness, fact, or a neutral view. A high-angle shot: to make the viewer feel more powerful than a character; to suggest detachment, vulnerability, or dependence. A low-angle shot: to emphasize a character's importance or power.
4. Panning: swiveling the camera left or right to follow a moving subject or survey the scene; crabbing: moving the camera right or left; tracking or dollying: moving the camera toward or away from a subject.
5. A long take is a single shot that lasts for a relatively long period of time. It can be used to make a film feel authentic.
6. Answers will vary. Possible shots: medium-long, tilted, two-shot, point-of-view, wide-angle, telephoto, zoom, following, whip-pan.

Check Yourself! Skill Check

1. $\tan 50° = 2640 \text{ ft}/d$
   $d = 2,640 \text{ ft}/\tan 50°$
   $d = 2,640/0.8391$
   $d = 3,146$ feet or 0.6 miles
2. At 10 feet from the squid, they would need an 84° view to see the creature from side to side—too wide for the typical standard lens. A wide angle lens would be best. At 15 feet, they could probably use a standard lens (62° field of vision).

Director's View

Expedition Overview

Challenge
Behind every great film are cinematographers and people maneuvering cameras. In this project, join the director and camera crew working on a scene from one of your favorite movies. Use your own artistic vision and geometry skills to tell them how to shoot the scene!

Objectives
- To learn the geometry involved in camera work and film production
- To learn how to communicate technical information about camera angles
- To understand how to use circle and angle relationships to solve field-of-vision problems

Project Activities
Before You Go
- Camera Geometry—Field of Vision
- Cinematography 101

Off You Go
- Activity 1: Scene Analysis
- Activity 2: Set Map

Expedition Tool
Shot Sheet

Other Materials Needed
- paper
- graph paper
- pencils
- ruler
- protractor
- scientific calculator
- video equipment such as camera, tripod, DVD, DVD player (optional)

Lingo to Learn—Terms to Know
- arc
- central angle
- chord
- circle
- field of vision
- inscribed angle
- major arc
- minor arc
- perpendicular

Expeditions in Your Classroom: Geometry

Director's View

Expedition Overview

Helpful Web Resources

- The Daily Script—Movie Scripts and Movie Screenplays
  www.dailyscript.com

- Exposure—Acting with a Pencil: Storyboarding Your Movie (via Edison High School's CIBACS program)
  www.cibacs.org/teacherpages/mwhitmore/downloads/pdf/sb/Storyboarding.pdf
  (Includes storyboard templates.)

- How Stuff Works—How Cameras Work
  http://electronics.howstuffworks.com/camera.htm

- How Stuff Works—How Digital Cameras Work
  www.howstuffworks.com/digital-camera.htm

- MediaKnowAll: Describing Shots
  www.mediaknowall.com/camangles.html

- Scripts on the Net—Movie Scripts & Screenplays Online
  www.roteirodecinema.com.br/scripts/

- The University of Wales—The "Grammar" of Television and Film
  www.aber.ac.uk/media/Documents/short/gramtv.html

- The Weekly Script
  www.weeklyscript.com

- Videomaker: What's Your Angle?
  www.videomaker.com/article/8723

- Wikipedia—Cinematic Techniques
  http://en.wikipedia.org/wiki/List_of_film_techniques

- Yale Film Studies—Film Analysis Guide
  http://classes.yale.edu/film-analysis
  (Click on the "Cinematography" link.)

Director's View

Before You Go

Camera Geometry—Field of Vision

> **Goals:** To learn how to "see" shots and angles as a camera does; to use geometry to solve field-of-vision problems
>
> **Materials:** scientific calculator, video camera (optional)

Did you know that humans have a wide field of view? A human can see a full 180°. Some birds can see a complete 360°. A standard camera lens "sees" a more limited view (50° to 60°).

Directions

1. If you have a video camera, experiment with its field of vision. Do not use zoom features.

2. Try to figure out how much you can "see" depending on where you stand. Think about where you would stand to have certain objects or areas "fill the picture."

3. With a few bits of information, you can precisely calculate ideal camera distances or determine how much the camera will see if positioned at a particular distance.

   Example: A camera has a 60° view. Where would a cameraperson stand if he or she wants a 30-foot house to just fill the screen? (see diagram on following page.)

   The left and right ends of the house are points *A* and *B*. They form the chord of a circle.

   The cameraperson could be standing within a 60° inscribed angle of the circle (the camera's view).

   The cameraperson should stand directly in front of the house. This forms an isosceles triangle that can be bisected to create right triangles.

   Calculate the unknown distance using tangent and known measurements of the right triangle:

   $\tan 30° = $ opposite/adjacent

   $\tan 30° = 15$ feet/d (distance)

   $d = 15$ feet/0.5774

   $d = 26$ feet

(continued)

Director's View

Before You Go

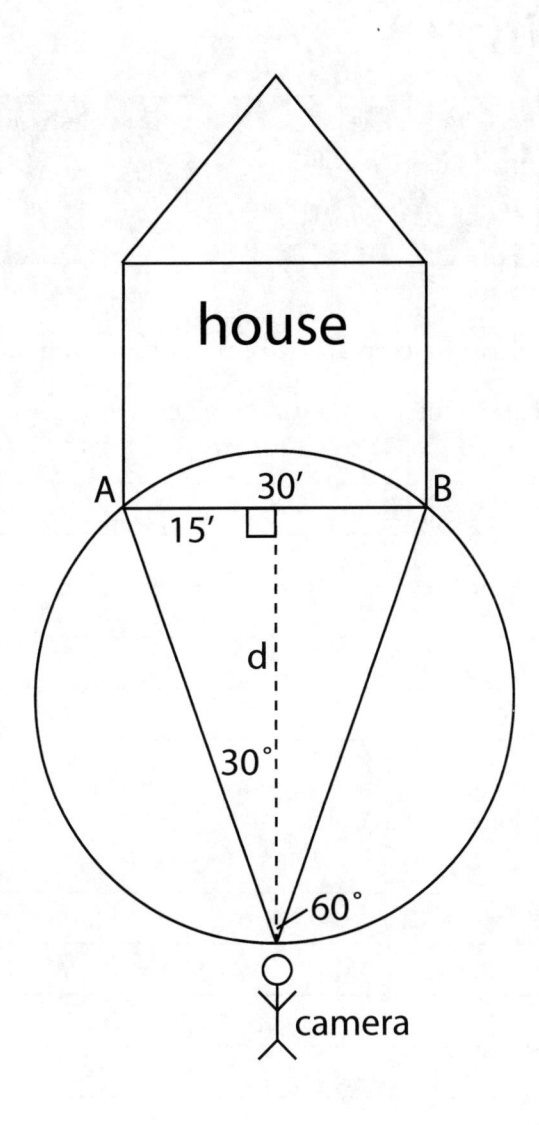

4. You are shooting a scene in which a car is the main focus. You want one shot of the car—and just the car—as viewed from the side. The length of the car is 175 inches. Your camera lens has a 58° view. How far away should you stand? Draw a diagram to illustrate your answer on a separate sheet of paper.

Director's View

Before You Go

Cinematography 101

> **Goal:** To explore the art of cinematography and the importance of camera angles
>
> **Materials:** computer with Internet access

Directions

Use your Helpful Web Resources to learn about camera angles and answer the questions below.

1. What are the three main elements of a camera shot?

2. Why might you use a long shot? A close-up?

3. Why might you use an eye-level shot? A high-angle shot? A low-angle shot?

(continued)

Director's View

Before You Go

4. What is the difference between panning, crabbing, and tracking (dollying)?

5. What is a long take?

6. Name five other camera shots or angles and describe them.

Director's View

Off You Go

Activity 1: Scene Analysis

Goal:	To analyze a scene from the script of a favorite film to set up camera angles
Materials:	computer with Internet access, TV/DVD player (optional)
Tools:	Shot Sheet

Directions

1. Use your Helpful Web Resources to find a screenplay and scene you want to use. Make sure that you choose a scene that has good potential for camera action (angular geometry).

2. Print out and read the section of the script that contains your scene or scenes (if one scene is too short).

3. Highlight anything that suggests a camera angle, view, or camera motion. If possible, watch the scene and take notes on camera action.

4. Reread the scene carefully. Brainstorm how you would film it. Assume that you have plenty of camera equipment and will use at least two different cameras. Think about the types of shots you would take and the number of shots you would need.

5. Make notes in your script. Use actual shot terminology to describe angles.

6. Using the Shot Sheet to guide you, create a list of the shots you want to take. Identify at least three shots. They may come from one camera or three; however, the camera must change position or view.

7. *Optional:* Create a rough storyboard of two or three main shots you want to capture.

   - Create your own storyboard template or download one from the storyboard site listed in your Helpful Web Resources.
   - You do not need to draw the entire scene. Choose moments where the view or action changes significantly—where you know you want to use a different camera angle.
   - Don't worry about artistic skill. Use stick figures and basic shapes to represent your ideas. The goal is to give someone looking at the storyboard a basic idea of the camera angles involved (wide shot, close-up, and so forth).
   - Leave enough space underneath each frame for one to two lines of text. Use this space to label the shot by number and summarize the camera action. For example: *Shot 2: Cut to medium range view of interior of bank. Zoom to show close-up of clock.*

Director's View

Expedition Tool

Shot Sheet

Date: _____

Producer (your name): _____

Film name: _____

Scene: _____

Provide at least three distinct shots for your scene.

Reel/shot number	Angle (high, medium, or low)	Shot type (abbreviate)	Shot description	Vectors/motion (zoom, pan, dolly, track, etc.)
1/Opening shot				

Available Camera Equipment

Here are the cameras available to you and their lens specifications. You must use at least two cameras during your scene.

Camera 1: standard lens with a 60° field of vision

Camera 2: standard lens with a 60° field of vision

Camera 3: wide-angle lens with a 100° field of vision

Camera 4: telephoto lens with a 15° field of vision

Camera 5: fish-eye (super-wide) angle lens with 180° field of vision

Camera 6: overhead camera and camera jib, standard lens with a 60° field of vision

If you don't see something you need, requisition it (check with your teacher). You will need to find field-of-vision information if the camera uses a lens other than those listed above.

Director's View

Off You Go

Activity 2: Set Map

> **Goal:** To create a set map that shows camera positions, angles, and action
>
> **Materials:** paper, pencil, colored pencils, ruler, protractor

Directions

1. Create a map (a floor plan or ground plan) of your set or location as it would look at the start of your scene—your opening shot.

Set Map Criteria

❑ Draw your map to scale or represent proportion as accurately as you can.

❑ Include any major features or objects that will appear in the scene (a desk, a car, a medical examiner's table, etc.).

❑ Provide dimensions for the set and any major objects or features.

❑ Indicate actors' marks—locations where actors will stand during the scene. You can mark each with an *X*.

❑ Provide the height of each actor. You can find this information or make it up.

Tip: Make a very rough sketch of your set first and work out your camera calculations outlined in numbers 2 and 3 below. Create your final map only after you are certain of your work.

2. Mark the starting location of each camera you plan to use in the scene.

   • Use field-of-vision information and other set and actor dimensions to calculate the location. Make sure the action captured by the camera "fills the picture."
   • Use pencil to illustrate each camera's initial field of vision.

3. Show the maximum range of motion of each camera during the shoot. For example:

   • Say camera 1 will pan 30° to the left for one shot and 40° to the right several shots later. It won't move beyond that range for the rest of the scene. Draw arrows to indicate left and right motion. Provide degree information and dotted lines or shading to show the new field of vision.
   • Say camera 2 will advance 10 feet. Label the starting point *A* and the new position *B*. Draw a line, label the distance, and use dotted lines or shading to show the new field of vision.

4. Review your completed map and turn it in to your teacher.

Director's View

Check Yourself!

Skill Check

1. You are on location filming a television commercial. You need to set up a long shot with a 100°-angle lens that will slowly pan the horizon to show New Mexico's beautiful landscape. You want the camera to stop with the biggest mesa (1 mile wide) in full view, then zoom in for a close-up of a climber halfway to the top guzzling down your client's energy drink product.

   How far away would you position the camera so that you can pan and capture the mesa view the way you want? Show your work.

2. A special-effects team is working on a sequence for a new 3-D film. They have an enormous automated squid that is 30 feet tall and 18 feet wide. While watching the finished movie, the audience will feel as if the squid is leaping from the screen to swallow them whole. The sound stage is tight on space. The camera can shoot the scene from 12 to 18 feet away. What is the minimum field of vision the camera should have? Show your work.

Expeditions in Your Classroom: Geometry

Director's View

Check Yourself!

Self-Assessment and Reflection
Project Management

Before You Go

- ❑ I understand the concept of field of vision (field of view) and how to use geometric principles to calculate camera position when given field of vision and other dimensions of set items.
- ❑ I'm honestly not sure I understand the math involved in the project and have asked my teacher for additional help.
- ❑ I researched camera angles. I can name and describe at least five different examples.

Off You Go

- ❑ I found a good scene for Activity 1: Scene Analysis. I printed the script for the scene and reviewed it to get camera-action ideas.
- ❑ I made a list of shots I would use to film the scene. My list includes at least three different shots.
- ❑ My set map is legible and well-labeled. It clearly shows key set dimensions, the starting location of cameras, and the maximum range of motion of each.

Do You Know?

- ❑ I can define the Lingo to Learn vocabulary terms for this project and give an example of each.
- ❑ I completed the Skill Check problems and carefully reviewed problems I answered incorrectly.

Director's View

Check Yourself!

Reflection

1. What were the most challenging aspects of this project for you and why?

   _____
   _____
   _____
   _____
   _____
   _____
   _____
   _____

2. What skills did this project help you develop?

   _____
   _____
   _____
   _____
   _____
   _____
   _____
   _____

3. If you did this project again, what might you do differently and why?

   _____
   _____
   _____
   _____
   _____
   _____
   _____
   _____

This Is Air Traffic Control

Overview
Students create air-traffic-control scenarios to practice their geometric flight and landing skills.

Time
Total time: 8 to 10 hours

- Before You Go—Air Traffic in 3-D: one 55-minute class period
- Before You Go—Air Traffic Control School: one 55-minute class period and 30 to 60 minutes of homework
- Before You Go—Landing Math: one 55-minute class period
- Activity 1—Where in the World Is Your Airport?: 90 to 120 minutes of homework
- Activity 2—Inbound Flights: one 55-minute class period and 30 to 60 minutes of homework
- Activity 3—Smooth Landings: one to two 55-minute class periods and/or homework assignments

Skill Focus
- apply three-dimensional visualization and graphing using Cartesian and polar systems
- learn how to find angle and side measurements using tangents (right triangle trigonometry)

Prior Knowledge
- coordinate systems/Cartesian graphs
- familiarity with tangents and use of scientific calculators
- Some experience with polar coordinates is helpful.

Team Formation
Students work individually or in pairs.

Lingo to Learn—Terms to Know
- **Cartesian coordinates:** coordinates where each point on a plane is determined using x-, y-, and z-axis values
- **coordinate system:** method of locating points or positions by assigning numbers to them; a system where every point on a plane is associated with a set of numbers

This Is Air Traffic Control

- **polar coordinates:** two-dimensional coordinates where each point on a plane is determined by an angle (θ or t = angular coordinate) and a distance (r = radial coordinate)
- **tangent:** the trigonometric ratio of the length of the side opposite an angle to the length of the side adjacent to the angle

Suggested Steps

Preparation

- Make five paper airplanes. Hang the paper airplanes from the ceiling of the classroom. Vary the height and location of planes, but have them all "flying" toward a location you select as the "airport"—a spot on the floor marked as X with masking tape. Be sure that students can access the planes so that they can collect location coordinates (measurements).
- Do a web search using the key words *runway map*. Find and print two or three examples of airplane runway maps.

Day 1

1. Provide an overview of the project and review project materials.

2. Facilitate Before You Go: Air Traffic in 3-D. Create groups to collect information on all planes, or ask one group to collect information on one plane on behalf of the class and report measurements. Write scale information on the board.

3. Before number 4, explain the difference between a Cartesian graph and a polar graph and show examples. Plot plane locations on a polar graph together as a class.

4. Review Before You Go: Landing Math. If students need more practice, ask them to help create scenarios (various angles, distances, altitudes). Do problems in class and assign as homework. Note: You might also save this activity for day 3 of the project (there are no project activities for class since students are working on runway maps at home).

5. Assign Before You Go: Air Traffic Control School as homework.

Homework

Have students read about air traffic control and finish Before You Go: Air Traffic Control School.

This Is Air Traffic Control

Day 2

1. Explain Activity 1: Where in the World Is Your Airport?
2. Review runway map criteria and show sample maps. Ask for observations. Ask the following questions:
   - How are runways laid out (are they parallel, across one another, etc.)?
   - What are the directions of the runways?
   - If you can see them, what numbers are used for runways? (Numbers refer to the direction and location of the runway.)
   - Where is the tower located?
3. Assign runway map due date.

Homework

Have students work on runway maps.

Day 3 (optional)

If not assigned earlier, have students complete Before You Go: Landing Math.

Day 4 (Runway Map Due Date)

1. Display runway maps around the room.
2. Call upon various students to explain their drawings, highlighting any geometry. Discuss as a class.
3. Provide an overview of Activity 2: Inbound Flights.
4. Assign only the En Route Flight Graph for now.
5. Review the Flight Progress Strip and Approach Plan. Remind students that their goal is to come up with reasonable, justifiable estimates based on average flight information provided.

Homework

Have students work on inbound flight progress strips and graph.

This Is Air Traffic Control

Day 5

1. Review student plane locations and information.

2. Call upon students to explain how they determined location, speed, altitude, and so forth. Discuss any revisions they had to make and why.

3. Review the Queued Flights Graph.

4. Give an example of instructions for queued flights (see the sample Flight Progress Strip and Approach Plan in the answer key). Students are to assume that speed remains constant for this activity. Note that this may make for a few unrealistic flight patterns. As an added challenge, students may adjust airspeed.

Homework

Have students continue their work on patterns and instructions for queued flights.

Day 6

1. Have students continue their work on queued flights or review.

2. Ask various students to explain their easiest and most challenging pattern examples.

3. Call upon students to explain their calculations.

Day 7

1. Explain Activity 3: Smooth Landings.

2. Review landing guidelines, as well as pattern legs if needed.

3. Give students an opportunity to work out at least one set of instructions during class.

4. Ask for one or two examples, and review calculations.

5. Assign the due date for Activity 3.

Homework

Have students work on final landing instructions.

This Is Air Traffic Control

Final Landings Due Date
1. Ask each student to present one or two examples of plane flight patterns.
2. As a class, look for and discuss overall patterns.
3. Given the original fix point of each plane, ask students what was needed to "land" the plane.
4. Discuss which planes landed safely, which did not, and why.
5. Ask if students notice any patterns (for example, pilots often double the length of the runway to estimate the length of their landing approach).

Final Day
1. Have students complete the Skill Check problems.
2. Check and review answers.
3. Have students complete the Self-Assessment and Reflection worksheet and submit it (optional).

Project Management Tips and Notes

- *Important:* There is no single correct answer. A student has a variety of options for "landing" a plane.
- Emphasize that students can choose any strategy that makes sense, as long as they can explain their assumptions and rationale and show correct calculations.
- For each assignment, give examples using the five paper planes you hung up in the classroom. For example, choose one plane and give an easy "queue fix point"—ask students to calculate the distance (using their scale grid) and the angle of descent.
- To shorten the project to two to four days: Assign the three Before You Go activities. However, skip Activity 1, and revise the flight plan activities as follows: Have each student hang one paper airplane in the classroom (pointing in any direction for a bit of extra directional fun). Indicate the "airport" and use masking tape to mark runways. Provide runway measurements using the same scale used for Before You Go: Air Traffic in 3-D ($1/4$ foot = 5 nautical miles, 1 nautical mile = 6,076 feet). Give students 5 to 8 minutes to identify and decorate their planes (adding the airline name, colors, and logo). Assign the runway and order each student should use for "landing." Students then complete a Flight Progress Strip and Approach Plan for their planes, with queuing and landing instructions. Queue "fix point" will depend on the landing order you assigned and safe separation from other planes. Students can "fly" (relocate) their planes into the correct queue positions.

Expeditions in Your Classroom: Geometry

This Is Air Traffic Control

Suggested Assessment
Use the Geometry Project Assessment Rubric or the following point system:

Class participation	10 points
Runway map	15 points
En Route Flight Progress Strips and Graph	15 points
Queued Flights Graph and instructions	30 points
Landing instructions	30 points

Extension Activities
- Try NASA's Smart Skies LineUp with Math simulator.
  http://smartskies.nasa.gov/lineup/index.html
- Ask a pilot to visit the classroom to explain takeoff and landing patterns.
- Take a tour of a local airfield.

Common Core State Standards Connection
High School
Geometry: Modeling with Geometry

G-MG.3. Apply geometric methods to solve design problems (e.g., designing an object or structure to satisfy physical constraints or minimize cost; working with typographic grid systems based on ratios).★

Geometry: Similarity, Right Triangles, and Trigonometry

G-SRT.6. Understand that by similarity, side ratios in right triangles are properties of the angles in the triangle, leading to definitions of trigonometric ratios for acute angles.

Answer Key
Before You Go: Air Traffic Control School

1. preflight, takeoff, departure, en route, descent, approach, landing

2. The pilot sends flight plan information to the control tower. A controller reviews it and enters it into an FAA computer, which produces a flight progress strip.

3. These are predetermined areas for ascent or descent activity. Controllers guide planes to these corridors for takeoffs or landings.

Expeditions in Your Classroom: Geometry

This Is Air Traffic Control

4. This refers to a plane's airspace—the vertical and horizontal zone around the craft that should be clear of other airplanes and so forth. It keeps planes at a safe distance from one another, and keeps planes from the effects of another plane's turbulent wake. Vertical separation below 29,000 feet altitude: 1,000 feet (305 meters); above 29,000 feet altitude: 2,000 feet (610 meters). Horizontal separation: 5 miles.

5. Descent begins when a plane is about 150 miles from its destination. The controller directs all planes flying to the destination to lower altitudes and merges them into a single-file line or queue, or places them in a holding pattern until the airport can handle the arrival. Approach refers to the plane's final approach to the airport before taking turns to align with the runway. An approach coordinator handles the initial stage, starting 50 miles out, and then turns the plane over to a local controller in the airport tower 10 miles out. The local controller handles the landing.

Before You Go: Landing Math

1. a. $\tan 3° = 425 \text{ feet}/x$

    $x = 425 \text{ feet}/\tan 3°$

    $x = 425/0.052$

    $x = 8{,}173 \text{ feet}$

    $x = 1.55 \text{ miles}$

    b. $\tan 3° = y/3 \text{ miles}$

    $\tan 3° = y/15{,}840 \text{ feet}$

    $y = \tan 3° \times 15{,}840 \text{ feet}$

    $y = 0.052 \times 15{,}840 \text{ feet}$

    $y = 824 \text{ feet}$

2. a. $\tan x = 250 \text{ feet}/2 \text{ nautical miles}$

    $\tan x = 250 \text{ feet}/12{,}152 \text{ feet}$

    $\tan x = 0.02$

    $x = 1.1°$

    b. $\tan 9° = 400 \text{ feet}/x$

    $x = 400 \text{ feet}/\tan 9°$

    $x = 400 \text{ feet}/0.158$

    $x = 2{,}531 \text{ feet or } 0.4 \text{ nautical mile}$

    c. $\tan 3° = 330 \text{ feet}/x$

    $x = 330 \text{ feet}/\tan 3°$

    $x = 330 \text{ feet}/0.052$

    $x = 6{,}346 \text{ feet}$

This Is Air Traffic Control

3. First calculate landing distance required (x) to maintain a 3° descent.

   $\tan 3° = 400 \text{ feet}/x$

   $x = 400/\tan 3°$

   $x = 400/0.052$

   $x = 7{,}692$ feet or 1.27 nautical miles

   Next, calculate the time it took to travel that distance (given that speed).

   $r = d/t$

   120 knots = 1.27 nautical miles/t

   $t = 1.27$ nautical miles/120 knots

   $t = 0.01$

Convert to minutes: $t = 0.01 \times 60 = 0.6$ minutes to descend 400 feet

   Finally, divide the change in elevation by the time.

   $r = d/t$

   $r = 400$ feet/0.6 minutes = 667 feet/minute

This Is Air Traffic Control

Expedition Tool: Flight Progress Strip and Approach Plan

Sample flight plan and instructions:

Controller name: 3D Sami Brown

Airport: Pineapple Bay Island International Jetport

Part I: Progress Strip—En Route Fix Point

Airline and flight number ZoomAir889	Boeing 747	Flight route (departure and destination points) Boston to Pineapple Bay Island nonstop	Beacon Code 2345
Current heading and coordinates Altitude: 30,000 feet Distance to airport: 150 nautical miles Heading: 90° (due east)	Airspeed: 500 knots	Current time: Estimated time of arrival:	

Part II: Flight Plan to Queue Position

Queue fix point:

Altitude: 15,000 feet

Distance from airport: 50 nautical miles

Speed: 500 knots (Students may keep constant for this leg or change.)

Pilot instructions: Make 90° turn onto heading 180° (due south). Maintain current speed but descend to altitude 15,000 feet over distance of 100 nautical miles.

Angle of descent: 14°

Rate of descent: 1,250 feet/minute

Expeditions in Your Classroom: Geometry

This Is Air Traffic Control

Part III: Flight Plan for Landing

Target point (position)	Instructions to point
Queue fix point Altitude: 15,000 feet Heading: 180° Distance to airport: 10 nautical miles	
Downwind leg fix point Altitude: 1,000 feet Heading: 225° Distance to airport: 10 nautical miles	Heading: Turn right 45° and adjust course to heading 225° (southwest) Angle of descent: 30° Distance to point: 40 nautical miles Average speed: 500 knots Average rate of descent: 2,916 feet/minute
Base leg fix point Altitude: 750 feet Heading: 315° Distance to airport: 5 nautical miles	Heading: Turn right 90° and adjust course to heading 315° (northwest) Angle of descent (from downwind midpoint): 1° Leg distance: 5 nautical miles Average speed: 350 knots Average rate of descent: 595 feet/minute
Final approach fix point: Heading: 55° Altitude: 500 feet Distance to airport: 2 nautical miles	Heading: Turn right 90° and adjust course to heading to 55° (northeast) Angle of descent: 1° Leg distance: 3 nautical miles Average speed: 200 knots Average rate of descent: 277 feet/minute
Runway touchdown point Heading: 145° Touchdown/airport: elevation 25 feet	Heading: Turn right 90° and adjust course to heading 145° for landing. Angle of descent: 3° Leg distance: 9135 feet/1.5 nautical miles Average speed: 150 knots Average rate of descent: 791 feet/minute

Expeditions in Your Classroom: Geometry © Walch Education

This Is Air Traffic Control

This plane did not land safely! It missed the runway by 0.5 mile (it landed before the runway). Distance plane began final approach: 2 nautical miles from airport. Distance needed to land: 1.5 nautical miles.

Other observations about the flight plan:

- The angle of descent used to position for the turn into the downwind leg seems very steep. The descent is for the downwind and base legs are very flat. It might have been better to even out descent during the initial approach legs.
- This is a great lesson in working the problem backwards, as a pilot would. You know you need a 3° angle of descent for the final leg (landing). How far away would the pilot need to be in order to maintain that angle and land in the touchdown zone? What would the altitude be at that point?

Check Yourself! Skill Check (sample answers)

1.

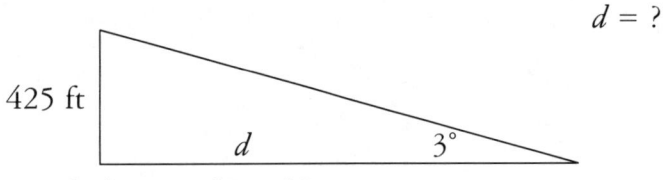

425 ft, d, 3°, d = ?

Angle/distance of Final Leg
tan 3° = 425 ft/d
d = 425 ft/tan 3°
d = 425 ft/.052
d = 8,173 ft

2. $\tan 3° = \dfrac{425}{x}$ x = 8173 or 1.35 nautical miles

$r = d/t$
150 knots = 1.35 nm/t
t = 1.35 nautical miles/150 knots
t = 0.009
t = 0.009 × 60 = 0.54 minutes to descend 425 feet

$r = d/t$
r = 425/0.54 minutes
r = 787 feet/minute

This Is Air Traffic Control

Expedition Overview

Challenge
You are an air traffic controller with a swarm of commercial airliners headed in your direction. Your mission is to guide each flight into your airport for a smooth touchdown. Inbound Flight 407 awaits your instructions! Will the plane make the runway?

Objectives
- To practice three-dimensional visualization and plot three-dimensional coordinates
- To understand the difference between Cartesian and polar coordinates
- To study the geometry of landing a plane; to use basic trigonometry to calculate angle, distance, and altitude measurements
- To apply geometric principles and concepts to real-world situations

Project Activities
Before You Go
- Air Traffic in 3-D
- Air Traffic Control School
- Landing Math

Off You Go
- Activity 1: Where in the World Is Your Airport?
- Activity 2: Inbound Flights
- Activity 3: Smooth Landings

Expedition Tool
Flight Progress Strip and Approach Plan

Other Materials Needed
- graph paper
- paper
- colored pencils
- tape measure, ruler, or yardstick
- scientific calculator
- trigonometric table

Lingo to Learn—Terms to Know
- Cartesian coordinate
- coordinate system
- polar coordinate
- tangent

Expeditions in Your Classroom: Geometry

This Is Air Traffic Control

Expedition Overview

Helpful Web Resources

- Federal Aviation Administration Academy
  www.faa.gov/about/office_org/headquarters_offices/arc/programs/academy

- How Stuff Works—How Air Traffic Control Works
  http://travel.howstuffworks.com/air-traffic-control.htm

- NASA Smart Skies—Air Traffic Control Simulator
  www.atcsim.nasa.gov

- NASA Smart Skies—LineUp with Math Simulator
  http://smartskies.nasa.gov/lineup/index.html

- National Air Traffic Controllers Association—Gate to Gate Career Guidance Package
  www.natca.org/ULWSiteResources/natcaweb/Resources/file/About%20NATCA/FAA_Career_Guidance_Pkg.pdf

- National Air Traffic Controllers Association—Gate to Gate Multimedia
  http://www.natca.org/career_day_classroom_materials.aspx?zone=Career%20Day/%20Classroom%20Materials&pID=708#p708

- The Math Forum—Ask Dr. Math: Coordinate Systems, Longitude, Latitude
  http://mathforum.org/library/drmath/view/56443.html

This Is Air Traffic Control

Before You Go

Air Traffic in 3-D

> **Goal:** To describe the location of an object or a point in space
>
> **Materials:** ruler, yardstick, or tape measure; graph paper

Your classroom is an air traffic zone. Note the five paper airplanes hanging from the ceiling. Determine the polar directions and the location of the destination airport!

Directions

1. Form groups according to your teacher's instructions. Identify the location (coordinates) of each plane and collect measurements. Record location data for all five planes in a table.

2. Using graph paper, create a three-dimensional graph that represents the locations of all five planes. You determine the origin of the graph. Be sure to label each axis and each plane and indicate scale. Make the following assumptions:

   - classroom height: $1/4$ foot = 1,000 feet
   - classroom length and depth: $1/4$ foot = 5 nautical miles
   - airport elevation: 500 feet above sea level

3. Calculate the distance and altitude between planes. Add this information to your table. If the planes continue to fly in the same direction, at the same speed, would any collide?

   _____
   _____
   _____
   _____

4. As a class, plot plane locations again—this time on a two-dimensional polar grid.

This Is Air Traffic Control

Before You Go

Air Traffic Control School

> **Goal:** To research air traffic control procedures, the stages of a flight, and how flight path information is conveyed
>
> **Materials:** computer with Internet access

Directions
Use your Helpful Web Resources to answer the questions below.

1. What are the stages of a flight? _____
   _____

2. How is flight plan information conveyed throughout the flight?
   _____
   _____

3. What are ascent/departure and descent/approach corridors?
   _____
   _____
   _____

4. What is "safe separation"? Why is it important? Typically, what is considered safe?
   _____
   _____
   _____
   _____

5. Explain the difference between descent and approach.
   _____
   _____
   _____
   _____

This Is Air Traffic Control

Before You Go

Landing Math

> **Goal:** To understand the angles involved in landing a plane
>
> **Materials:** scientific calculator, trigonometric table

Directions

1. The standard approach angle pilots use to descend during the final leg of a flight is 3°. This keeps landings smooth and safe. Sometimes the pilot must identify how high the plane should be to begin descent (altitude) or how long the approach segment should be (length of approach). Both scenarios are shown below. For a, calculate the distance. For b, calculate the altitude. Show your work below or on another sheet of paper.

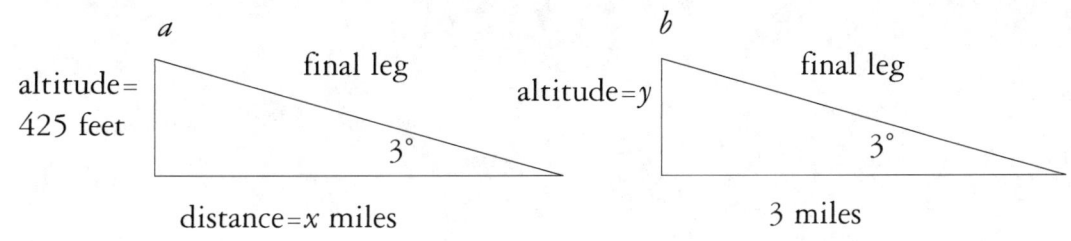

a
altitude = 425 feet
final leg
3°
distance = *x* miles

b
altitude = *y*
final leg
3°
3 miles

When landing, pilots don't usually fly straight for the runway, especially at large airports. The approach path is a maneuver with angled turns. The standard approach geometry includes the following legs:

- the downwind leg, which runs parallel to the runway in the opposite direction the plane will eventually land (Descent can begin at the midpoint of this leg or during the next leg.)
- a 90° turn into the base leg
- a 90° turn into the final leg, during which the plane aligns with the centerline of the runway and makes the landing

The pattern is sometimes called the "trombone" because controllers can adjust the length of the downwind leg.

(continued)

This Is Air Traffic Control

Before You Go

2. A pilot wants to turn into the downwind leg at an altitude of 1,000 feet. After 2 nautical miles, at the midpoint of the leg, she plans to begin descent (the rate of descent is constant). She wants to begin the base leg at 750 feet, use a 9° angle of descent, and turn for her final approach at an altitude of 350 feet. Assume no wind. Airport elevation is 20 feet. Answer the following questions. Show your work below or on a separate sheet of paper.

   a. What is the angle of descent for the second portion of the downwind leg?

   b. How long is the base leg?

   c. How long is the final leg?

3. A pilot is in his final approach, flying 120 knots at 500 feet. If he wants to maintain a 3° descent to land (airport elevation 100 feet), what will his rate of descent be? Give rate in feet per minute. Remember that the formula to calculate rate is distance/time, or $r = d/t$. Show your work below or on a separate sheet of paper.

Expeditions in Your Classroom: Geometry

This Is Air Traffic Control

Off You Go

Activity 1: Where in the World Is Your Airport?

> **Goal:** To create an airport runway map
> **Materials:** computer with Internet access, paper, colored pencils, ruler, protractor

Directions

1. Use the Internet to find examples of airport runway maps.

2. Determine the location of your airport. Choose any location (for example, a city near you or a location you would love to visit).

3. Draw a runway map for your airport. You can base your map on an actual commercial airport or "build" a new one. Unless your teacher instructs otherwise, you do not need to draw precisely to scale; however, your map should give a relatively accurate view. You may need to do additional research to help you make your measurements realistic.

4. Post completed airport maps around the class and compare drawings.

5. Hold onto your map. You will need it for another activity.

Runway Map Criteria

Your airport must accommodate commercial flights. Your runway map should include the following:

- ❑ one or more numbered runways with width and length dimensions
- ❑ the center line and "touchdown zone" marked on each runway
- ❑ taxiways
- ❑ terminal building
- ❑ one or more gates and ramps (gate areas)
- ❑ air traffic control tower
- ❑ airport elevation
- ❑ airport latitude and longitude information
- ❑ compass or direction indicator (north, east, south, west)
- ❑ labels for all map elements

This Is Air Traffic Control

Off You Go

Activity 2: Inbound Flights

Goal:	To plot the current positions of flights en route
Materials:	paper or graph paper, colored pencils
Tools:	Flight Progress Strip and Approach Plan (one per flight)

Directions

1. You are an air traffic controller at the airport you created. Four planes are en route to your destination. Each is coming from a different direction (north, south, east, west). Each is within 300 miles of the airport but not closer than 150 miles. Identify the current position (fix) of each plane and create a flight progress strip for it. No two planes should have the same flight path, speed, or altitude. Invent the information, but make it realistic. For example, a typical cruising altitude is between 30,000 and 43,000 feet; a typical cruising speed is between 430 and 530 miles per hour.

2. Plot the location of your planes on a three-dimensional En Route Flight Graph.

3. Label each plane as follows:

   - Label the *x*-axis, *y*-axis, and *z*-axis and indicate scale.
   - Label direction (north, east, south, and west).
   - Include the airport.
   - Include a table with coordinate information.

4. The first plane is 50 miles away. You need to direct the planes into single lines and into approach corridors so they can begin descent. How would you queue them up? Remember, your airport has more than one runway, so decide where to send each plane.

   a. Create a Queued Flights Graph—a two-dimensional polar grid—that shows the order and position of planes in queue.

      - Label the *x*-axis and *y*-axis and indicate scale.
      - Label each plane and give coordinates relative to the airport (altitude, distance).
      - Put a circle around each plane to indicate the separation zone.
      - Include the airport.

   b. On the Flight Progress Strip and Approach Plan, write the instructions you "gave" each plane. Assume wind, weather, and so forth are not factors.

5. Draw a diagram that shows the path each plane took to get into position.

   - Indicate changes in course heading, target altitude, and, if appropriate, speed.
   - Don't forget to maintain safe separation between planes.

Expeditions in Your Classroom: Geometry © Walch Education

This Is Air Traffic Control

Expedition Tool

Flight Progress Strip and Approach Plan

Complete one progress strip for each incoming flight. Remember to invent information but be realistic.

Controller name: _____

Airport: _____

Part I: Progress Strip—En Route Fix Point

Airline and flight number	Boeing 747	Flight route (departure and destination points)	Beacon code 2345
Current heading and coordinates Altitude: Distance to airport (nautical miles): Heading:	Airspeed (knots)	Current time: Estimated time of arrival:	

Attach your En Route Flight Graph to this worksheet.

Now record approach and landing instructions for this plane below. Give fix point information in altitude and nautical miles (from airport).

Part II: Flight Plan to Queue Position

Queue fix point: _____

Pilot instructions: _____

_____

_____

_____

_____

_____

_____

Show your work, including diagrams, on a separate sheet of paper.

(continued)

This Is Air Traffic Control

Expedition Tool

Part III: Flight Plan for Landing

Target point (position)	Instructions to point
Queue fix point Altitude: Heading: Distance to airport:	
Downwind leg fix point Altitude: Heading: Distance to airport:	Heading: Angle of descent: Distance to point: Average speed: Average rate of descent:
Base leg fix point Altitude: Heading: Distance to airport:	Heading: Angle of descent: Leg distance: Average speed: Average rate of descent:
Final approach fix point Heading: Altitude: Distance to airport:	Heading: Angle of descent: Leg distance: Average speed: Average rate of descent:
Runway touchdown point Heading: Touchdown/airport: Elevation:	Heading: Angle of descent: Leg distance: Average speed: Average rate of descent:

Show your work, including diagrams, on a separate sheet of paper.

(continued)

This Is Air Traffic Control

Expedition Tool

Did your plane land on target? If not, what could you have done differently?

_____
_____
_____
_____
_____
_____
_____
_____
_____

Record any other observations about the plan for this flight. What would you change or adjust?

_____
_____
_____
_____
_____
_____
_____
_____
_____

This Is Air Traffic Control

Expedition Tool

Activity 3: Smooth Landings

Goal:	To describe and illustrate landing instructions for your planes
Materials:	paper or graph paper, colored pencils, scientific calculator, trigonometric table
Tools:	Flight Progress Strip and Approach Plan

Directions

1. Guide your pilots in for a safe landing by giving them approach instructions. Record the instructions on the Flight Progress Strip and Approach Plan for each flight.

2. Use the landing strips on your runway map as final targets. Your instructions should include the following:

   - information needed to position for the downwind leg of the landing pattern (downwind leg fix point)
   - vectors (course headings) for each leg of the landing pattern—the downwind, base, and final legs
   - diagrams that show the final flight pattern of each plane and your work (measurements and labels)

3. Provide instructions for each leg using the following guidelines:

   - Enter your downwind leg at any reasonable angle (45° would be typical).
   - With the exception of the downwind leg, maintain a constant angle of descent for each leg. For the final leg, use the standard 3° approach angle.
   - Your final leg should begin 500 feet above the airport, a typical landing entry altitude.
   - Assume a constant rate of speed for landing.
   - Your landing speed is 150 knots.
   - Wind and weather are not factors.
   - Identify your runway touchdown point.

   Reminders:
   - The downwind leg runs parallel to the runway in the opposite direction the plane will land.
   - Your base leg is a 90° angle from the downwind leg.
   - Your final leg is a 90° angle from the base leg.
   - 1 knot = 1 nautical mile per hour = 6,076 feet per hour
   - 1 mph = 1 mile per hour = 5,280 feet per hour

   Tip: Determine how far you need to be from the end of the runway when you turn into your final approach.

   (continued)

This Is Air Traffic Control

Off You Go

Answer the following questions. Be prepared to share examples and explain your calculations in class!

4. Given starting fix points of planes, what did flight patterns look like for the last part of the journey? Any wild, crazy, or unrealistic flights (steep descent angles, sharp turns, and so forth)?

   _____
   _____
   _____

5. What was your strategy for devising a reasonable approach and landing?

   _____
   _____
   _____
   _____

6. Would you change anything about your flight instructions?

   _____
   _____
   _____
   _____

7. In the end, how did airport runways accommodate incoming flights?

   _____
   _____
   _____
   _____

8. What were the most challenging aspects of calculating flight and landing information?

   _____
   _____
   _____
   _____

Expeditions in Your Classroom: Geometry © Walch Education

This Is Air Traffic Control

Check Yourself!

Skill Check

A pilot is approaching an island resort with a plane full of happy vacationers. The island is known for its small runway—one that ends rather abruptly at the foot of a large volcanic mountain. The runway is 4,000 feet long. Ideal touchdown is between 500 and 800 feet. The plane is currently flying at an altitude of 1,000 feet and moving into position for a final turn so that it can enter the final leg at an altitude of 450 feet. Airport elevation is 25 feet above sea level.

1. How far away should the pilot begin his final approach in order to maintain a smooth 3° angle of descent, and land within the touchdown zone?

2. In his final approach, the pilot is flying 150 knots at 450 feet. If he wants to maintain a 3° descent to land, what will his rate of descent be? Give the rate in feet per minute.

This Is Air Traffic Control

Check Yourself!

Self-Assessment and Reflection
Project Management

Before You Go

- ❑ I participated fully in the activity Before You Go: Air Traffic in 3-D. The class's graph accurately reflects the locations of planes in the room. It is clearly labeled and includes measurements for distances between planes.
- ❑ I understand how to represent an airplane's coordinates using Cartesian and polar grids (both in two dimensions and three dimensions).
- ❑ I know how to calculate angle, distance, and altitude measurements associated with a plane's descent or landing.
- ❑ I'm honestly not sure I understand the math involved in the project and have asked my teacher for additional help.
- ❑ I read about air traffic control and provided clear, detailed answers to the questions in Before You Go: Air Traffic Control School.

Off You Go

- ❑ I reviewed the activities and materials for this project and understood the products I had to create.
- ❑ My runway map meets all required criteria and is realistic (used Internet examples).
- ❑ I completed a flight progress strip for each of my planes. I was creative but realistic. If needed, I did research to identify a reasonable measure for flight information.
- ❑ My En Route Flight Graph matches the flight strip information I provided.
- ❑ My Queued Flights Graph accurately represents the positions of planes based on the flight instructions I gave.
- ❑ My instructions for queued flights include header, altitude, and, if appropriate, speed information. My planes are queued with safe distances between them. I included drawings that clearly illustrate the flight path each plane took to get into queue position.
- ❑ My final landing instructions are detailed and include altitude, angle of descent, speed, and distance information. I included detailed drawings that show the landing pattern and measurements.

Do You Know?

- ❑ I can define the Lingo to Learn vocabulary terms for this project and give an example of each.
- ❑ I completed the Skill Check problems and carefully reviewed problems I answered incorrectly.

This Is Air Traffic Control

Check Yourself!

Reflection

1. What were the most challenging aspects of this project for you and why?

2. What skills did this project help you develop?

3. If you did this project again, what might you do differently and why?

The Great Geometry Race

Overview
Students create problems based on geometry in and around the school to help them review course topics.

Time
Total time: 3 to 4 hours

- Before You Go—Stop and Stumpers: one to two 55-minute class periods
- Activity—The Great Geometry Race: one to two 55-minute class periods

Skill Focus
This is a geometry review—use it at the end of the year, at the end of a semester, or to support a specific unit or topic.

Prior Knowledge
geometry course material

Team Formation
Students work in teams of two to four students for five to six stops within one 55-minute class period. Adjust team sizes, the number of teams, the number of questions per stop, or the number of stops to fit the time allotted.

Suggested Steps
Preparation

- Explain the project to school administrators and colleagues. Gain permission to use locations for the race.
- Create examples of the types of questions you expect students to generate. Use a fairly obvious location—one that students are unlikely to select (such as your classroom).
- Create a stop of your own—one that all teams must complete.
- Identify specific topics if you want to provide students with a focus.
- Consider matching teams to a topic. Use this strategy if you want the race to provide a balanced review of topics studied. Otherwise, students will be inspired by their surroundings and may gravitate toward topics with which they are more comfortable.

The Great Geometry Race

Day 1

1. Provide an overview of the race and review project tasks. Assign teams.

2. Explain Before You Go: Stop and Stumpers. Review stop and question criteria. Tell students to write down any additional criteria or ground rules.

3. Assign specific topics or skills stops should cover. Alternatively, let teams choose from a hat or a cup.

4. Inform teams that they have 40 minutes to scout out a stop location. Remind teams to stay focused and organized, and provide a completed, legible copy of their Stop and Stumpers activity and answer key by the end of class.

5. Send students off to work.

6. Direct students to return when finished, or five minutes before the end of class.

7. Collect Stop and Stumpers sheets.

8. Review Stop and Stumpers sheets. Flag them as "approved" or "needs more work." Make sure answers are correct. You may need to go to the location to verify answers. If needed, ask for edits or revisions. For example, a team may not have developed the type or level of question required by the topic you gave them.

9. Inventory materials needed. Place items at the stops (for example, items unique to each stop), or have teams pick items up at the start of the race if items are needed at multiple stops.

Day 2

1. Return Before You Go: Stop and Stumpers. Have students make corrections or revisions. Provide time for teams to rewrite their sheets as needed.

2. Collect final Stop and Stumpers sheets and answer keys.

3. Ask students if they have the items they need for the race (tape measures). Assign gathering supplies as homework.

4. Copy Stop and Stumpers sheets and answer keys separately. Copy the back side if needed. Include your stop if you created one. Note that each team will have one page they can ignore (the stop they created).

5. Provide the teams with one set of Stop and Stumpers sheets and answer keys.

The Great Geometry Race

Day 3 (Race Day)

1. Gather materials. Place items at the appropriate stops or on desks in the classroom.

2. Review race rules and scoring.

3. Direct students to stand against the back wall of the room with their team.

4. Give one person on each team a set of Stop and Stumpers sheets—facedown (no looking). Remind teams to list team member names on the front page.

5. Start the race and note the time. Suggest that teams take a few minutes to review their Stop and Stumpers sheets, strategize, and collect materials.

6. Record finish time at the moment a team hands you their completed packet.

7. Ask teams to sit together and wait for other teams to finish.

8. If time allows, continue on to scoring and answer review. You will need at least 15 to 20 minutes. Otherwise, score and announce race results the following day.

9. To score in class, return packets to teams along with answers. Tell teams to score their results as you review each stop and question. Students must use the honor system.

10. Have each team report the total number of incorrect and unanswered questions. Have them add the corresponding penalty seconds to their score and record their overall results on the first sheet. Invite scores to be reported and declare the winner. Collect packets to verify scores. As race judge, you may toss any question at your discretion (for example, if information was misleading and no team answered or all teams answered incorrectly).

Day 4

1. If you scored results yourself, do not announce them yet.

2. If students are to score, follow steps 9 and 10 above.

3. Hand sheets back along with an answer key. Review each stop and question.

4. Ask teams to explain various answers (correct and incorrect).

5. After you complete the answer review, announce the results.

Expeditions in Your Classroom: Geometry

The Great Geometry Race

Final Day

1. Have students complete the Skill Check problems.
2. Check and review answers.
3. Have students complete the Self-Assessment and Reflection worksheet and submit it (optional).

Project Management Tips and Notes

- Focus or broaden the content and/or challenge of the race by assigning a topic or type of problem to each team.
- Add a stop or ask students to add questions if you think the race isn't covering material you want reviewed.
- To simplify and shorten the project, you can provide all stops and problems.
- As written, students base their race problems on geometry they find around the building. If needed, you can add a stop with general questions that do not relate to location features.

Suggested Assessment

Use the Geometry Project Assessment Rubric or the following point system:

Team and class participation	20 points
Stop and stumpers	40 points
Race answers/score	35 points
Project self-assessment	5 points

Common Core State Standards Connection

This Expedition is meant as a review; therefore, the standards addressed will depend upon the project the students design.

The Great Geometry Race

Expedition Overview

Challenge
Put your knowledge and skills to the test. Develop one "stop" on a geometry race course that will wind through your school. At each stop, teams will try to answer stumpers—problems you base on geometric objects or features visible at the stop location.

Objectives
- To review your geometry knowledge as you race through a course

Project Activities
Before You Go
- Stop and Stumpers

Off You Go
- Activity: The Great Geometry Race

Expedition Tool
- Stop and Stumpers sheet

Other Materials Needed
- paper
- pencils
- calculator
- ruler
- tape measures (one per team)
- protractor
- stopwatch

Helpful Web Resources
- edHelper.com—High School Geometry: Geometry Worksheets
  www.edhelper.com/geometry_highschool.htm

- Math.com—Homework Help: Geometry
  www.math.com/homeworkhelp/Geometry.html

The Great Geometry Race

Before You Go

Stop and Stumpers

> **Goal:** To create the questions or "stumpers" for one stop of the geometry review race
>
> **Materials:** graph paper, pencil, ruler, tape measure (one per team)
>
> **Tools:** Stop and Stumpers sheet

Directions

1. As a team, scout out locations in the school with excellent geometry-problem potential. Look for interesting two- and three-dimensional shapes and patterns. For example, you might check out the shapes of trays and tables in the cafeteria, or stop by the gym to examine basketball court or bleacher geometry.

2. Choose a final location and develop a geometry "stop" for the race.

Geometry Stop Criteria

- ❑ Teams must solve at least one geometry problem at your stop.
- ❑ Problems must be based on concepts and skills covered in class.
- ❑ You may have more than one problem or question, but no more than four. Problems can be related (for example, have teams find at least three isosceles triangles and calculate the area of each) or not.
- ❑ At least one problem must involve a geometric calculation.
- ❑ Teams must be able to solve your problem safely.
- ❑ Make the stop challenging!

---

3. Use the Stop and Stumpers sheet to provide information about your stop. Write each problem teams must solve.

4. Attach a separate sheet that provides the answer to each problem. Show your work.

5. Give your Stop and Stumpers sheet and answer key to your teacher for review. Make any revisions or clarifications requested.

The Great Geometry Race

Expedition Tool

Stop and Stumpers

Team name: _____

Team members: _____

Stop name: _____

Stop location: _____

Materials other teams need (graph paper, ruler, and so forth):
_____
_____

Problems

Write or type each problem you want teams to solve at this stop. Number each problem. Provide enough space for the answer (leave space for a diagram, a figure, or a grid if needed). Use a separate sheet of paper if necessary. Attach a separate page with your answer key.

The Great Geometry Race

Off You Go

Activity: The Great Geometry Race

Goal:	To complete the race in the fastest time possible with the most correct answers
Materials:	paper, pencil, calculator, protractor, ruler, measuring tape
Tools:	Stop and Stumpers sheets from other teams

Directions

1. It's race day! Review the rules and wait for the signal.

   Geometry Race Rules

   ❑ Your team must stay together. You must all go to the same stop at the same time.
   ❑ Your team can decide the best strategy for each stop.
   ❑ You must visit each stop and answer each question—or pay the penalty.
   ❑ You must put your answers on the appropriate Stop and Stumpers sheet. You may use scrap and graph paper to work out solutions.
   ❑ You will collect the materials you need at the start of the race or find them at the stop. If your team forgets an item, the entire team must return to the classroom to get it.
   ❑ Return to the classroom when done. Your finish time is the moment you hand your completed Stop and Stumpers packet to your teacher. Be sure to list team names on the front page.

2. Get your team's Stop and Stumpers sheets for all stops.

3. Collect the materials you will need before you race on.

4. Listen for the signal from your teacher to start the race.

   Geometry Race Scoring

   ❑ 20 seconds will be added to your team time for each incorrect answer.
   ❑ 60 seconds will be added to your team time for each unanswered question.
   ❑ The team with the fastest time, including penalties, wins. In the event of a tie, the team with the most correct answers wins!

The Great Geometry Race

Check Yourself!

Self-Assessment and Reflection
Project Management

Before You Go

- ❏ I reviewed the terms and topics my teacher wants to cover with the geometry race.
- ❏ I carefully reviewed the criteria for a race stop and helped my team scout out locations.
- ❏ I made a significant contribution to the set of geometry review questions we developed for our stop.
- ❏ I double-checked the list of materials a team would need for our stop to be sure we didn't miss anything.
- ❏ I helped make any revisions to our questions that my teacher requested.
- ❏ I'm honestly not sure I understand the project and have asked my teacher for additional help.

Off You Go

- ❏ I participated fully in the race and helped solve problems at every stop.
- ❏ I helped review answers at each stop to check for mistakes.
- ❏ I made notes about questions I didn't understand or couldn't help answer.
- ❏ I helped keep my team focused and organized during the race.
- ❏ During the answer review, I made notes about skills or topics I need to practice—and I did review them.

Reflection

1. What were the most challenging aspects of this project for you and why?

   _____
   _____
   _____
   _____
   _____

(continued)

The Great Geometry Race

Check Yourself!

2. What skills did this project help you develop?

   _____
   _____
   _____
   _____
   _____
   _____

3. If you did this project again, what might you do differently and why?

   _____
   _____
   _____
   _____
   _____
   _____